The Art of Thinking Clearly

The Art of Thinking Clearly

ROLF DOBELLI

Translated by Nicky Griffin

HARPER

www.harpercollins.com

HarperCollins books may be purchased for educational, business, or sales promotional use. For information, please e-mail the Special Markets Department at SPsales@harpercollins.com.

FIRST EDITION

Designed by Michael P. Correy

Library of Congress Cataloging-in-Publication Data

Dobelli, Rolf.
  [Kunst des klaren Denkens. English]
  The art of thinking clearly / Rolf Dobelli; translated by Nicky Griffin.—First edition.
      p.   cm.
  Translation of the author's Die Kunst des klaren Denkens, published by Hanser in 2012.
  ISBN: 978-0-06-221968-8
  1. Reasoning (Psychology). 2. Errors—Psychological aspects. 3. Decision making. 4. Cognition. I. Title.
  BF442.D63   2013
  153.4'2—dc23   2013003934

13 14 15 16 17   OV/RRD   10 9 8 7 6 5 4 3 2 1

FOR SABINE

CONTENTS

Contents

Contents

Contents

Contents

Contents

INTRODUCTION

In the fall of 2004, a European media mogul invited me to Munich to partake in what was described as an "informal exchange of intellectuals." I had never considered myself an "intellectual"—I had studied business, which made me quite the opposite, really—but I had also written two literary novels and that, I guessed, must have qualified me for such an invitation.

Nassim Nicholas Taleb was sitting at the table. At that time, he was an obscure Wall Street trader with a penchant for philosophy. I was introduced to him as an authority on the English and Scottish Enlightenment, particularly the philosophy of David Hume. Obviously I had been mixed up with someone else. Stunned, I nevertheless flashed a hesitant smile around the room and let the resulting silence act as proof of my philosophical prowess. Right away, Taleb pulled over a free chair and patted the seat. I sat down. After a cursory exchange about Hume, the conversation mercifully shifted to Wall Street. We marveled at the systematic errors in decision making CEOs and business leaders make—ourselves included. We chatted about the fact that unexpected events seem much more likely in retrospect. We chuckled about why it is that investors cannot part with their shares when they drop below acquisition price.

Following the event, Taleb sent me pages from his manuscript, a gem of a book, which I commented on and partly criticized. These went on to form part of his international best seller, *The Black Swan*. The book catapulted Taleb into the intellectual all-star league. Meanwhile, my appetite whetted, I began to devour books and articles written by cognitive and social scientists on topics such as "heuristics and biases," and I also increased my e-mail conversations with a large number researchers and started to visit their labs. By 2009, I realized that, alongside my job as a novelist, I had become a student of social and cognitive psychology.

The failure to think clearly, or what experts call a "cognitive error," is a systematic deviation from logic—from optimal, rational, reasonable thought and behavior. By "systematic," I mean that these are not just occasional errors in judgment but rather routine mistakes, barriers to logic we stumble over time and again, repeating patterns through generations and through the centuries. For example, it is much more common that we overestimate our knowledge than we underestimate it. Similarly, the danger of losing something stimulates us much more than the prospect of making a similar gain. In the presence of other people we tend to adjust our behavior to theirs, not the opposite. Anecdotes make us overlook the statistical distribution (base rate) behind it, not the other way round. The errors we make follow the same pattern over and over again, piling up in one specific, predictable corner like dirty laundry, while the other corner remains relatively clean (i.e., they pile up in the "overconfidence corner," not the "underconfidence corner").

To avoid frivolous gambles with the wealth I had accumulated over the course of my literary career, I began to put together a list of these systematic cognitive errors, complete with notes and personal anecdotes—with no intention of ever publishing them. The list was originally designed to be used by me alone. Some of these thinking errors have been known for centuries; others have been discovered in the last few years. Some come with two or three names attached to them. I chose the terms most widely used. Soon I realized that such a compilation of pitfalls was not only useful for making investing decisions but also for business and personal matters. Once I had prepared the list, I felt calmer and more levelheaded. I began to recognize my own errors sooner and was able to change course before any lasting damage was done. And, for the first time in my life, I was able to recognize when others might be in the thrall of these very same systematic errors. Armed with my list, I could now resist their pull—and perhaps even gain an upper hand in my dealings. I now had categories, terms, and explanations with which to ward off the specter of irrationality. Since Benjamin Franklin's kite-flying days, thunder and lightning have not grown less frequent, powerful, or loud—but they have become less worrisome. This is exactly how I feel about my own irrationality now.

Friends soon learned of my compendium and showed interest. This led to a weekly newspaper column in Germany, Holland, and Switzerland, countless presentations (mostly to medical doctors, investors, board members, CEOs, and government officials), and eventually to this book.

Please keep in mind three things as you peruse these pages: First, the list of fallacies in this book is not complete. Undoubtedly new ones will be discovered. Second, the majority of these errors are related to one another. This should come as no surprise. After all, all brain regions are linked. Neural projections travel from region to region in the brain; no area functions independently. Third, I am primarily a novelist and an entrepreneur, not a social scientist; I don't have my own lab where I can conduct experiments on cognitive errors, nor do I have a staff of researchers I can dispatch to scout for behavioral errors. In writing this book, I think of myself as a translator whose job is to interpret and synthesize what I've read and learned—to put it in terms others can understand. My great respect goes to the researchers who, in recent decades, have uncovered these behavioral and cognitive errors. The success of this book is fundamentally a tribute to their research. I am enormously indebted to them.

This is not a how-to book. You won't find "seven steps to an error-free life" here. Cognitive errors are far too engrained to rid ourselves of them completely. Silencing them would require superhuman willpower, but that isn't even a worthy goal. Not all cognitive errors are toxic, and some are even necessary for leading a good life. Although this book may not hold the key to happiness, at the very least it acts as insurance against too much self-induced unhappiness.

Indeed, my wish is quite simple: If we could learn to recognize and evade the biggest errors in thinking—in our private lives, at work, or in government—we might experience a leap in prosperity. We need no extra cunning, no new ideas, no unnecessary gadgets, no frantic hyperactivity—all we need is less irrationality.

The Art of
Thinking
Clearly

1

Why You Should Visit Cemeteries
Survivorship Bias

No matter where Rick looks, he sees rock stars. They appear on television, on the front pages of magazines, in concert programs, and at online fan sites. Their songs are unavoidable—in the mall, on his playlist, in the gym. The rock stars are everywhere. There are lots of them. And they are successful. Motivated by the stories of countless guitar heroes, Rick starts a band. Will he make it big? The probability lies a fraction above zero. Like so many others, he will most likely end up in the graveyard of failed musicians. This burial ground houses ten thousand times more musicians than the stage does, but no journalist is interested in failures—with the exception of fallen superstars. This makes the cemetery invisible to outsiders.

In daily life, because triumph is made more visible than failure, you systematically overestimate your chances of succeeding. As an outsider, you (like Rick) succumb to an illusion, and you mistake how minuscule the probability of success really is. Rick, like so many others, is a victim of *survivorship bias*.

Behind every popular author you can find a hundred other writers whose books will never sell. Behind them are another hundred who haven't found publishers. Behind them are yet another hundred whose unfinished manuscripts gather dust in drawers. And behind each one of these are a hundred people who dream of—one day—writing a book. You, however, hear of only the successful authors (these days, many of them self-published) and fail to recognize how unlikely literary success is. The same goes for photographers, entrepreneurs, artists, athletes, architects, Nobel Prize winners, television presenters, and beauty queens. The media is not interested in digging around in the graveyards of the unsuccessful. Nor is this its job. To elude the *survivorship bias*, you must do the digging yourself.

You will also come across *survivorship bias* when dealing with money and risk: Imagine that a friend founds a start-up. You belong to the circle of potential investors and you sense a real opportunity: This could be the next Google. Maybe you'll be lucky. But what is the reality? The most likely scenario is that the company will not even make it off the starting line. The second most likely outcome is that it will go bankrupt within three years. Of the companies that survive these first three years, most never grow to more than ten employees. So, should you never put your hard-earned money at risk? Not necessarily. But you should recognize that the *survivorship bias* is at work, distorting the probability of success like cut glass.

Take the Dow Jones Industrial Average index. It consists of out-and-out survivors. Failed and small businesses do not enter the stock market, and yet these represent the majority of business ventures. A stock index is not indicative of a country's economy. Similarly, the press does not report proportionately on all musicians. The vast number of books and coaches dealing with success should also you make skeptical: The unsuccessful don't write books or give lectures on their failures.

Survivorship bias can become especially pernicious when you become a member of the "winning" team. Even if your success stems from pure coincidence, you'll discover similarities with other winners and be tempted to mark these as "success factors." However, if you ever visit the graveyard of failed individuals and companies, you will realize that its tenants possessed many of the same traits that characterize your success.

If enough scientists examine a particular phenomenon, a few of these studies will deliver statistically significant results through pure coincidence—for example, the relationship between red wine consumption and high life expectancy. Such (false) studies immediately attain a high degree of popularity and attention. As a result, you will not read about the studies with the "boring" but correct results.

Survivorship bias means this: People systematically overestimate their chances of success. Guard against it by frequently visiting the graves of once-promising projects, investments, and careers. It is a sad walk but one that should clear your mind.

2

Does Harvard Make You Smarter?
Swimmer's Body Illusion

As essayist and trader Nassim Taleb resolved to do something about the stubborn extra pounds he'd been carrying, he contemplated taking up various sports. However, joggers seemed scrawny and unhappy, and body-builders looked broad and stupid, and tennis players? Oh, so upper-middle-class! Swimmers, though, appealed to him with their well-built, streamlined bodies. He decided to sign up at his local swimming pool and to train hard twice a week.

A short while later, he realized that he had succumbed to an illusion. Professional swimmers don't have perfect bodies because they train extensively. Rather, they are good swimmers because of their physiques. How their bodies are designed is a factor for selection and not the result of their activities. Similarly, female models advertise cosmetics and, thus, many female consumers believe that these products make you beautiful. But it is not the cosmetics that make these women model-like. Quite simply, the models are born attractive, and only for this reason are they candidates for cosmetics advertising. As with the swimmers' bodies, beauty is a factor for selection and not the result.

Whenever we confuse selection factors with results, we fall prey to what Taleb calls the *swimmer's body illusion*. Without this illusion, half of advertising campaigns would not work. But this bias has to do with more than just the pursuit of chiseled cheekbones and chests. For example, Harvard has the reputation of being a top university. Many highly successful people have studied there. Does this mean that Harvard is a good school? We don't know. Perhaps the school is terrible, and it simply recruits the brightest students around. I experienced this phenomenon at the University of St. Gallen in Switzerland. It is said to be one of the top ten business schools in Europe, but the lessons I received (albeit twenty-five years ago) were mediocre. Nevertheless, many of its graduates were successful. The reason behind this is unknown—perhaps it was due to the climate in the narrow valley or even the cafeteria food. Most probable, however, is the rigorous selection.

All over the world, MBA schools lure candidates with statistics regarding future income. This simple calculation is supposed to show that the horrendously high tuition fees pay for themselves over a short period of time. Many prospective students fall for this approach. I am not implying that the schools doctor the statistics, but still their statements must not be swallowed wholesale. Why? Because those who pursue an MBA are different from those who do not. The income gap between both groups stems from a multitude of reasons that have nothing to do with the MBA degree itself. Once again we see the *swimmer's body illusion* at work: the factor for selection confused with the result. So, if you are considering further study, do it for reasons other than a bigger paycheck.

When I ask happy people about the secret of their contentment, I often hear answers like "You have to see the glass half full rather than half empty." It is as if these individuals do not realize that they were born happy and now tend to see the positive in everything. They do not realize that cheerfulness—according to many studies, such as those conducted by Harvard's Dan Gilbert—is largely a personality trait that remains constant throughout life. Or, as social scientists David Lykken and Auke Tellegen starkly suggest, "trying to be happier is as futile as trying to be taller." Thus, the *swimmer's body illusion* is also a self-illusion. When these optimists write self-help books, the illusion can become treacherous. That's why it's important to give wide berth to tips and advice from self-help authors. For billions of people, these pieces of advice are unlikely to help. But because the unhappy don't write self-help books about their failures, this fact remains hidden.

In conclusion: Be wary when you are encouraged to strive for certain things—be it abs of steel, immaculate looks, a higher income, a long life, a particular demeanor, or happiness. You might fall prey to the *swimmer's body illusion*. Before you decide to take the plunge, look in the mirror—and be honest about what you see.

3

Why You See Shapes in the Clouds
Clustering Illusion

In 1957, Swedish opera singer Friedrich Jorgensen bought a tape player to record his vocals. When he listened back to the recording, he heard strange noises throughout, whispers that sounded like supernatural messages. A few years later, he recorded birdsong. This time, he heard the voice of his deceased mother in the background whispering to him: "Fried, my little Fried, can you hear me? It's Mammy." That did it. Jorgensen turned his life around and devoted himself to communicating with the deceased via tape recordings.

In 1994, Diane Duyser from Florida also had an otherworldly encounter. After biting into a slice of toast and placing it back down on the plate, she noticed the face of the Virgin Mary in it. Immediately, she stopped eating and stored the divine message (minus a bite) in a plastic container. In November 2004, she auctioned the still fairly well preserved snack on eBay. Her daily bread earned her $28,000.

In 1978, a woman from New Mexico had a similar experience. Her tortilla's blackened spots resembled Jesus's face. The press latched on to the story, and thousands of people flocked to New Mexico to see the savior in burrito form. Two years earlier, in 1976, the orbiter of the Viking spacecraft photographed a rock formation that, from high above, looked like a human face. The "face on Mars" made headlines around the world.

And you? Have you ever seen faces in the clouds or the outlines of animals in rocks? Of course. This is perfectly normal. The human brain seeks patterns and rules. In fact, it takes it one step further: If it finds no familiar patterns, it simply invents some. The more diffuse the signal, such as the background noise on the tape, the easier it is to find "hidden messages" in it. Twenty-five years after uncovering the "face on Mars," the Mars global surveyor sent back crisp, clear images of the rock formations: The captivating human face had dissolved into plain old scree.

These frothy examples make the *clustering illusion* seem innocuous; it is not. Consider the financial markets, which churn out floods of data every second. Grinning from ear to ear, a friend told me that he had discovered a pattern in the sea of data: "If you multiply the percentage change of the Dow Jones by the percentage change of the oil price, you get the move of the gold price in two days' time." In other words, if share prices and oil climb or fall in unison, gold will rise the day after tomorrow. His theory worked well for a few weeks, until he began to speculate with ever-larger sums and eventually squandered his savings. He had sensed a pattern where none existed.

oxxxoxxxoxxoooxooxxoo. Is this sequence random or planned? Psychology professor Thomas Gilovich interviewed hundreds of people for an answer. Most did not want to believe the sequence was arbitrary. They figured some law must govern the order of the letters. Wrong, explained Gilovich, and pointed to some dice: It is quite possible to roll the same number four times in a row, which mystifies many people. Apparently we have trouble accepting that such events can take place by chance.

During World War II, the Germans bombed London. Among other ammunition, they used V1 rockets, a kind of self-navigating drone. With each attack, the impact sites were carefully plotted on a map, terrifying Londoners: They thought they had discovered a pattern and developed theories about which parts of the city were the safest. However, after the war, statistical analysis confirmed that the distribution was totally random. Today it's clear why: The V1's navigation system was extremely inaccurate.

In conclusion: When it comes to pattern recognition, we are oversensitive. Regain your skepticism. If you think you have discovered a pattern, first consider it pure chance. If it seems too good to be true, find a mathematician and have the data tested statistically. And if the crispy parts of your pancake start to look a lot like Jesus's face, ask yourself: If he really wants to reveal himself, why doesn't he do it in Times Square or on CNN?

4

If Fifty Million People Say Something Foolish, It Is Still Foolish

Social Proof

You are on your way to a concert. At an intersection, you encounter a group of people, all staring at the sky. Without even thinking about it, you peer upward, too. Why? *Social proof.* In the middle of the concert, when the soloist is displaying absolute mastery, someone begins to clap and suddenly the whole room joins in. You do, too. Why? *Social proof.* After the concert you go to the coat check to pick up your coat. You watch how the people in front of you place a coin on a plate, even though, officially, the service is included in the ticket price. What do you do? You probably leave a tip as well.

Social proof, sometimes roughly termed the "herd instinct," dictates that individuals feel they are behaving correctly when they act the same as other people. In other words, the more people who follow a certain idea, the better (truer) we deem the idea to be. And the more people who display a certain behavior, the more appropriate this behavior is judged by others. This is, of course, absurd.

Social proof is the evil behind bubbles and stock market panic. It exists in fashion, management techniques, hobbies, religion, and diets. It can paralyze whole cultures, such as when sects commit collective suicide.

A simple experiment, carried out in the 1950s by legendary psychologist Solomon Asch, shows how peer pressure can warp common sense. A subject is shown a line drawn on paper, and next to it three lines—numbered 1, 2, and 3—one shorter, one longer, and one the same length as the original one. He or she must indicate which of the three lines corresponds to the original one. If the person is alone in the room, he gives correct answers because the task is really quite simple. Now five other people enter the room; they are all actors, which the subject does not know. One after another, they give wrong answers, saying "number 1," although it's very clear that number 3 is the correct answer. Then it is the subject's turn again. In one-third of cases, he will answer incorrectly to match the other people's responses.

Why do we act like this? Well, in the past, following others was a good survival strategy. Suppose that fifty thousand years ago you were traveling around the Serengeti with your hunter-gatherer friends, and suddenly they all bolted. What would you have done? Would you have stayed put, scratching your head, and weighing up whether what you were looking at was a lion or something that just looked like a lion but was in fact a harmless animal that could serve as a great protein source? No, you would have sprinted after your friends. Later on, when you were safe, you could have reflected on what had actually happened. Those who acted differently—and I am sure there were some—exited the gene pool. We are the direct heirs of those who copied the others' behavior. This pattern is so deeply rooted in us that we still use it today, even when it offers no survival advantage. Only a few cases come to mind where *social proof* is of value. For example, if you find yourself hungry in a foreign city and don't know a good restaurant, it makes sense to pick the one that's full of locals. In other words, you copy the locals' behavior.

Comedy and talk shows make use of *social proof* by inserting canned laughter at strategic spots, inciting the audience to laugh along. One of the most impressive, though troubling, cases of this phenomenon is the famous speech by Nazi propaganda minister Joseph Goebbels, delivered to a large audience in 1943. (See it for yourself on YouTube.) As the war went from bad to worse for Germany, he demanded to know: "Do you want total war? If necessary, do you want a war more total and radical than anything that we can even imagine today?" The crowd roared. If the attendees had been asked individually and anonymously, it is likely that nobody would have consented to this crazy proposal.

The advertising industry benefits greatly from our weakness for *social proof.* This works well when a situation is unclear (such as deciding among various car makes, cleaning products, beauty products, and so on, with no obvious advantages or disadvantages), and where people "like you and me" appear.

So be skeptical whenever a company claims its product is better because it is "the most popular." How is a product better simply because it sells the most units? And remember English novelist W. Somerset Maugham's wise words: "If fifty million people say something foolish, it is still foolish."

5

Why You Should Forget the Past

Sunk Cost Fallacy

T he film was dire. After an hour, I whispered to my wife: "Come on, let's go home." She replied: "No way. We're not throwing away thirty dollars." "That's no reason to stay," I protested. "The money's already gone. This is the *sunk cost fallacy* at work—a thinking error!" She glared at me as if she had just bitten off a piece of lemon. Okay, I sometimes go overboard on the subject, itself an error called *déformation professionnelle* (see chapter 92). I desperately tried to clarify the situation. "We have spent the thirty dollars regardless of whether we stay or leave, so this factor should not play a role in our decision." Needless to say, I gave in and sunk back down in my seat.

The next day, I sat in a marketing meeting. Our advertising campaign had been running for four months and had not met even one of its goals. I was in favor of scrapping it. The advertising manager resisted, saying: "But we've invested so much money in it. If we stop now, it'll all have been for nothing." Another victim of the *sunk cost fallacy*.

A friend struggled for years in a troubled relationship. His girlfriend cheated on him time and again. Each time, she came back repentant and begged for forgiveness. He explained it to me this way: "I've invested so much energy in the relationship, it would be wrong to throw it away." A classic case of the *sunk cost fallacy*.

The *sunk cost fallacy* is most dangerous when we have invested a lot of time, money, energy, or love in something. This investment becomes a reason to carry on, even if we are dealing with a lost cause. The more we invest, the greater the sunk costs are, and the greater the urge to continue becomes.

Investors frequently fall victim to the *sunk cost fallacy*. Often they base their trading decisions on acquisition prices. "I lost so much money with this stock, I can't sell it now," they say. This is irrational. The acquisition price should play no role. What counts is the stock's future performance (and the future performance of alternative investments). Ironically, the more money a share loses, the more investors tend to stick by it.

This irrational behavior is driven by a need for consistency. After all, consistency signifies credibility. We find contradictions abominable. If we decide to cancel a project halfway through, we create a contradiction: We admit that we once thought differently. Carrying on with a meaningless project delays this painful realization and keeps up appearances.

The Concorde is a prime example of a government deficit project. Even though both parties, Britain and France, had long realized that the supersonic aircraft business would never work, they continued to invest enormous sums of money in it— if only to save face. Abandoning the project would have been tantamount to admitting defeat. The *sunk cost fallacy* is therefore often referred to as the "Concorde effect." It leads to costly, even disastrous, errors of judgment. The Americans extended their involvement in the Vietnam War because of this. Their thinking: "We've already sacrificed so much for this war; it'd be a mistake to give up now."

"We've come this far . . ." "I've read so much of this book already . . ." "But I've spent two years doing this course . . ." If you recognize any of these thought patterns, it shows that the *sunk cost fallacy* is at work in a corner of your brain.

Of course, there may be good reasons to continue investing in something to finalize it. But beware of doing so for the wrong reasons, such as to justify non-recoverable investments. Rational decision making requires you to forget about the costs incurred to date. No matter how much you have already invested, only your assessment of the future costs and benefits counts.

6

Don't Accept Free Drinks
Reciprocity

Not so long ago, you may have come across disciples of the Hare Krishna sect floating around in saffron-colored robes as you hurried to catch a flight or a train to your destination. A member of the sect presented you with a small flower and a smile. If you're like most people, you took the flower simply to avoid seeming rude. If you tried to refuse, you would have heard a gentle "Take it, this is our gift to you." If you wanted to dispose of the flower in the next trash can, you found that there were already a few there. But that was not the end. Just as your bad conscience started to tug at you, another disciple of Krishna approached you again, this time asking for a donation. In many cases, this plea was successful—so pervasive that many airports banned the sect from the premises.

Psychologist Robert Cialdini can explain the success of this and other such campaigns. He has studied the phenomenon of *reciprocity* and has established that people have extreme difficulty being in another person's debt.

Many NGOs and philanthropic organizations use exactly the same techniques: First give, then take. Last week, a conservation organization sent me an envelope full of postcards featuring all sorts of idyllic landscapes. The accompanying letter assured me that the postcards were a gift to be kept, whether or not I decided to donate to their organization. Even though I understood the tactic, it took a little willpower and ruthlessness to throw them in the trash.

Unfortunately, this kind of gentle blackmail—you could also call it corruption—is widespread. A supplier of screws invites a potential customer to join him at a big sports game. A month later, it's time to order screws. The desire not to be in debt is so strong that the buyer gives in and places an order with his new friend.

It is also an ancient technique. We find *reciprocity* in all species whose food supplies are subject to high fluctuations. Suppose you are a hunter-gatherer. One day you are lucky and kill a deer. You can't possibly eat all of it in a day, and refrigerators are still a few centuries away. You decide to share the deer with the group, which ensures that you will benefit from others' spoils when your haul is less impressive. The bellies of your buddies serve as your refrigerator.

Reciprocity is a very useful survival strategy, a form of risk management. Without it, humanity—and countless species of animals—would be long extinct. It is at the core of cooperation between people (who are not related) and a necessary ingredient for economic growth and wealth creation. There would be no global economy without it—there would be no economy at all. That's the good side of reciprocity.

But there is also an ugly side of *reciprocity*: retaliation. Revenge breeds counter-revenge, and you soon find yourself in a full-scale war. Jesus preached that we should break this cycle by turning the other cheek, which proves very difficult to do, so compelling is the pull of *reciprocity* even when the stakes are far less high.

Several years ago, a couple invited my wife and me to dinner. We had known this couple casually for quite some time. They were nice but far from entertaining. We couldn't think of a good excuse to refuse, so we accepted. Things played out exactly as we had imagined: The dinner party was beyond tedious. Nevertheless, we felt obliged to invite them to our home a few months later. The constraint of *reciprocity* had now presented us with two wearisome evenings. And, lo and behold, a few weeks later, a follow-up invitation from them arrived. I wonder how many dinner parties have been endured in the name of *reciprocity*, even if the participants would have preferred to drop out of the vicious cycle years ago.

In much the same way, if someone approaches you in the supermarket, whether to offer you a taste of wine, a chunk of cheese, or a handful of olives, my best advice is to refuse their offer—unless you want to end up with a refrigerator full of stuff you don't even like.

7

Beware the "Special Case"
Confirmation Bias (Part 1)

Gil wants to lose weight. He selects a particular diet and checks his progress on the scale every morning. If he has lost weight, he pats himself on the back and considers the diet a success. If he has gained weight, he writes it off as a normal fluctuation and forgets about it. For months, he lives under the illusion that the diet is working, even though his weight remains constant. Gil is a victim of the *confirmation bias*—albeit a harmless form of it.

The *confirmation bias* is the mother of all misconceptions. It is the tendency to interpret new information so that it becomes compatible with our existing theories, beliefs, and convictions. In other words, we filter out any new information that contradicts our existing views ("disconfirming evidence"). This is a dangerous practice. "Facts do not cease to exist because they are ignored," said writer Aldous Huxley. However, we do exactly that, as super-investor Warren Buffett knows: "What the human being is best at doing is interpreting all new information so that their prior conclusions remain intact."

The *confirmation bias* is alive and well in the business world.

One example: An executive team decides on a new strategy. The team enthusiastically celebrates any sign that the strategy is a success. Everywhere the executives look, they see plenty of confirming evidence, while indications to the contrary remain unseen or are quickly dismissed as "exceptions" or "special cases." They have become blind to disconfirming evidence.

What can you do? If the word "exception" crops up, prick up your ears. Often it hides the presence of disconfirming evidence. It pays to listen to Charles Darwin: Since his youth, he set out to fight the *confirmation bias* systematically. Whenever observations contradicted his theory, he took them very seriously and noted them down immediately. He knew that the brain actively "forgets" disconfirming evidence after a short time. The more correct he judged his theory to be, the more actively he looked for contradictions.

The following experiment shows how much effort it takes to question your own theory. A professor presented his students with the number sequence 2–4–6. They had to calculate the underlying rule that the professor had written on the back of a sheet of paper. The students had to provide the next number in the sequence to which the professor would reply "fits the rule" or "does not fit the rule." The students could guess as many numbers as they wanted but could try only once to identify the rule. Most students suggested 8 as the next number, and the professor replied: "Fits the rule." To be sure, they tried 10, 12, and 14. The professor replied each time: "Fits the rule." The students concluded: "The rule is to add two to the last number." The professor shook his head: "That is not the rule."

One shrewd student tried a different approach. He tested out the number –2. The professor said: "Does not fit the rule."

"Seven?" he asked. "Fits the rule." The student tried all sorts of numbers: –24, 9, –43. Apparently he had an idea, and he was trying to find a flaw with it. Only when he could no longer find a counterexample, the student said: "The rule is this: The next number must be higher than the previous one." The professor turned over the sheet of paper, revealing those very words. What distinguished the resourceful student from the others? While the majority of students sought merely to confirm their theories, he tried to find fault with his, consciously looking for disconfirming evidence. You might think: "Good for him, but not the end of the world for the others." However, falling for the *confirmation bias* is not a petty intellectual offense. How it affects our lives will be revealed in the next chapter.

8

Murder Your Darlings

Confirmation Bias (Part 2)

In the previous chapter, we met the father of all fallacies, the *confirmation bias*. Here are a few examples of it: We are forced to establish beliefs about the world, our lives, the economy, investments, our careers, and more. We deal mostly in assumptions, and the more nebulous these are, the stronger the *confirmation bias*. Whether you go through life believing that "people are inherently good" or "people are inherently bad," you will find daily proof to support your case. Both parties, the philanthropists and the misanthropes, simply filter disconfirming evidence (evidence to the contrary) and focus on the do-gooders and dictators who support their worldviews.

Astrologers and economists operate on the same principle. They utter prophecies so vague that any event can substantiate them: "In the coming weeks you will experience sadness," or "In the medium term, the pressure on the dollar will increase." But what is the medium term? What will cause the dollar to depreciate? And depreciation measured against what—gold, yen, pesos, wheat, residential property in Manhattan, the average price of a hot dog?

Religious and philosophical beliefs represent an excellent breeding ground for the *confirmation bias*. Here, in soft, spongy terrain, it grows wild and free. For example, worshippers always find evidence for God's existence, even though he never shows himself overtly—except to illiterates in the desert and in isolated mountain villages. It is never to the masses in, say, Frankfurt or New York. Counterarguments are dismissed by the faithful, demonstrating just how powerful the *confirmation bias* is.

No professionals suffer more from the *confirmation bias* than business journalists. Often, they formulate an easy theory, pad it out with two or three pieces of "evidence," and call it a day. For example: "Google is so successful because the company nurtures a culture of creativity." Once this idea is on paper, the journalist corroborates it by mentioning a few other prosperous companies that foster ingenuity. Rarely does the writer seek out disconfirming evidence, which in this instance would be struggling businesses that live and breathe creativity or, conversely, flourishing firms that are utterly uncreative. Both groups have plenty of members, but the journalist simply ignores them. If he or she were to mention just one, the story line would be ruined.

Self-help and get-rich-quick books are further examples of blinkered storytelling. Their shrewd authors collect piles of proof to pump up the most banal of theories, such as "meditation is the key to happiness." Any reader seeking disconfirming evidence does so in vain: Nowhere in these books do we see people who lead fulfilled lives without meditation, or those who, despite meditation, are still sad.

The Internet is particularly fertile ground for the *confirmation bias*. To stay informed, we browse news sites and blogs, forgetting that our favored pages mirror our existing values, be they liberal, conservative, or somewhere in between. Moreover, a lot of sites now tailor content to personal interests and browsing history, causing new and divergent opinions to vanish from the radar altogether. We inevitably land in communities of like-minded people, further reinforcing our convictions—and the *confirmation bias.*

Literary critic Arthur Quiller-Couch had a memorable motto: "Murder your darlings." This was his advice to writers who struggled with cutting cherished but redundant sentences. Quiller-Couch's appeal is not just for hesitant hacks but for all of us who suffer from the deafening silence of assent. To fight against the *confirmation bias,* try writing down your beliefs—whether in terms of worldview, investments, marriage, health care, diet, career strategies—and set out to find disconfirming evidence. Axing beliefs that feel like old friends is hard work but imperative.

9

Don't Bow to Authority
Authority Bias

The first book of the Bible explains what happens when we disobey a great authority: We get ejected from paradise. This is also what less celestial authorities would have us believe—political pundits, scientists, doctors, CEOs, economists, government heads, sports commentators, consultants, and stock market gurus.

Authorities pose two main problems to clear thinking: First, their track records are often sobering. There are about one million trained economists on the planet, and not one of them could accurately predict the timing of the 2008 financial crisis (with the exception of Nouriel Roubini and Nassim Taleb), let alone how the collapse would play out, from the real estate bubble bursting to credit default swaps collapsing, right through to the full-blown economic crunch. Never has a group of experts failed so spectacularly. The story from the medical world is much the same: Up until 1900 it was discernibly wiser for patients to avoid doctor's visits; too often the "treatment" only worsened the illness, due to poor hygiene and folk practices such as bloodletting.

Psychologist Stanley Milgram demonstrated the *authority bias* most clearly in an experiment in 1961. His subjects were instructed to administer ever-increasing electrical shocks to a person sitting on the other side of a pane of glass. They were told to start with 15 volts, then 30 volts, 45 volts, and so on, until they reached the maximum—a lethal dose of 450 volts. In reality, no electrical current was actually flowing; Milgram used an actor to play the role of the victim, but those charged with administering the shocks didn't know that. The results were, well, shocking: As the person in the other room wailed and writhed in pain, and the subject administering the shock wanted to stop, the professor would say, "Keep going, the experiment depends on it." The majority of people continued with the electrocution. More than half of the participants went all the way up to the maximum voltage—out of sheer obedience to authority.

Over the past decade, airlines have also learned the dangers of the *authority bias*. In the old days, the captain was king. His commands were not to be doubted. If a copilot suspected an oversight, he wouldn't have dared to address it out of respect for—or fear of—his captain. Since this behavior was discovered, nearly every airline has instituted crew resource management (CRM), which coaches pilots and their crews to discuss any reservations they have openly and quickly. In other words: They carefully deprogram the *authority bias*. CRM has contributed more to flight safety in the past twenty years than have any technical advances.

Many companies are light-years from this sort of foresight. Especially at risk are firms with domineering CEOs, where employees are likely to keep their "lesser" opinions to themselves—much to the detriment of the business.

Authorities crave recognition and constantly find ways to reinforce their status. Doctors and researchers sport white coats. Bank directors don suits and ties. Kings wear crowns. Members of the military wield rank badges. Today, even more symbols and props are used to signal expertise: appearances on talk shows and on the covers of magazines, to book tours and Wikipedia entries. Authority changes much like fashion does, and society follows it just as much.

In conclusion: Whenever you are about to make a decision, think about which authority figures might be exerting an influence on your reasoning. And when you encounter one in the flesh, do your best to challenge him or her.

10

Leave Your Supermodel Friends at Home

Contrast Effect

In his book *Influence*, Robert Cialdini tells the story of two brothers, Sid and Harry, who ran a clothing store in 1930s America. Sid was in charge of sales and Harry led the tailoring department. Whenever Sid noticed that the customers who stood before the mirror really liked their suits, he became a little hard of hearing. He called to his brother: "Harry, how much for this suit?" Harry looked up from his cutting table and shouted back: "For that beautiful cotton suit, forty-two dollars." (At that time, it was a completely inflated price.) Sid pretended as if he hadn't understood: "How much?" Harry yelled again: "Forty-two dollars!" Sid then turned to his customer and reported: "He says twenty-two dollars." At this point, the customer would have quickly put the money on the table and hastened from the store with the suit before poor Sid noticed his "mistake."

Maybe you know the following experiment from your school days: Take two buckets. Fill the first with lukewarm water and the second with ice water. Dip your right hand into the ice water for one minute. Then put both hands into the lukewarm water. What do you notice? The lukewarm water feels as it should to the left hand and piping hot to the right hand.

Both of these stories epitomize the *contrast effect*: We judge something to be beautiful, expensive, or large if we have something ugly, cheap, or small in front of us. We have difficulty with absolute judgments.

The *contrast effect* is a common misconception. You order leather seats for your new car because, compared to the $60,000 price tag on the car, $3,000 seems a pittance. All industries that offer upgrade options exploit this illusion.

The *contrast effect* is at work in other places, too. Experiments show that people are willing to walk an extra ten minutes to save $10 on food. But those same people wouldn't dream of walking ten minutes to save $10 on a $1,000 suit. An irrational move because ten minutes is ten minutes, and $10 is $10. Logically, you should walk back in both cases or not at all.

Without the *contrast effect*, the discount business would be completely untenable. A product that has been reduced from $100 to $70 seems a better value than a product that has always cost $70. The starting price should play no role. The other day an investor told me: "The share is a great value because it's 50 percent below the peak price." I shook my head. A share price is never "low" or "high." It is what it is, and the only thing that matters is whether it goes up or down from that point.

When we encounter contrasts, we react like birds to a gunshot. We jump up and get moving. Our weak spot: We don't notice small, gradual changes. A magician can make your watch vanish because, when he presses on one part of your body, you don't notice the lighter touch on your wrist as he relieves you of your Rolex. Similarly, we fail to notice how our money disappears. It constantly loses its value, but we do not notice because inflation happens over time. If it were imposed on us in the form of a brutal tax (and basically that's what it is), we would be outraged.

The *contrast effect* can ruin your whole life: A charming woman marries a fairly average man. But because her parents were awful people, the ordinary man appears to be a prince. One final thought: Bombarded by advertisements featuring supermodels, we now perceive beautiful people as only moderately attractive. If you are seeking a partner, never go out in the company of your supermodel friends. People will find you less attractive than you really are. Go alone or, better yet, take two ugly friends.

11

Why We Prefer a Wrong Map to None at All
Availability Bias

Smoking can't be that bad for you: My grandfather smoked three packs of cigarettes a day and lived to be more than a hundred." Or: "Manhattan is really safe. I know someone who lives in the middle of the Village and he never locks his door. Not even when he goes on vacation, and his apartment has never been broken into." We use statements like these to try to prove something, but they actually prove nothing at all. When we speak like this, we succumb to the *availability bias*.

Are there more English words that start with a *k* or more words with *k* as its third letter? Answer: More than twice as many English words have *k* in the third position than start with a *k*. Why do most people believe the opposite is true? Because we can think of words beginning with a *k* more quickly. They are more available to our memory.

The *availability bias* says this: We create a picture of the world using the examples that most easily come to mind. This is idiotic, of course, because in reality, things don't happen more frequently just because we can conceive of them more easily.

Thanks to the *availability bias*, we travel through life with an incorrect risk map in our heads. Thus, we systematically overestimate the risk of being the victims of a plane crash, a car accident, or a murder. And we underestimate the risk of dying from less spectacular means, such as diabetes or stomach cancer. The chances of bomb attacks are much rarer than we think, and the chances of suffering depression are much higher. We attach too much likelihood to spectacular, flashy, or loud outcomes. Anything silent or invisible we downgrade in our minds. Our brains imagine showstopping outcomes more readily than mundane ones. We think dramatically, not quantitatively.

Doctors often fall victim to the *availability bias*. They have their favorite treatments, which they use for all possible cases. More appropriate treatments may exist, but these are in the recesses of the doctors' minds. Consequently, they practice what they know. Consultants are no better. If they come across an entirely new case, they do not throw up their hands and sigh: "I really don't know what to tell you." Instead, they turn to one of their more familiar methods, whether or not it is ideal.

If something is repeated often enough, it gets stored at the forefront of our minds. It doesn't even have to be true. How often did the Nazi leaders have to repeat the term "the Jewish question" before the masses began to believe that it was a serious problem? You simply have to utter the words "UFO," "life energy," or "karma" enough times before people start to credit them.

The *availability bias* has an established seat at the corporate board's table, too. Board members discuss what management has submitted—usually quarterly figures—instead of more important things, such as a clever move by the competition, a slump in employee motivation, or an unexpected change in customer behavior. They tend not to discuss what's not on the agenda. In addition, people prefer information that is easy to obtain, be it economic data or recipes. They make decisions based on this information rather than on more relevant but harder-to-obtain information—often with disastrous results. For example, we have known for ten years that the so-called Black-Scholes formula for the pricing of derivative financial products does not work. But we don't have another solution, so we carry on with an incorrect tool. It is as if you were in a foreign city without a map, and then pulled out one for your hometown and simply used that. We prefer *wrong* information to *no* information. Thus, the *availability bias* has presented the banks with billions in losses.

What was it that Frank Sinatra sang—something about loving the girl I'm near when I'm not near the girl I love? A perfect example of the *availability bias*. Fend it off by spending time with people who think differently than you do—people whose experiences and expertise are different from yours. We require others' input to overcome the *availability bias*.

12

Why "No Pain, No Gain" Should Set Alarm Bells Ringing

The It'll-Get-Worse-Before-It-Gets-Better Fallacy

A few years ago, I was on vacation in Corsica and fell sick. The symptoms were new to me, and the pain was growing by the day. Eventually I decided to seek help at a local clinic. A young doctor began to inspect me, prodding my stomach, gripping my shoulders and knees, and then poking each vertebra. I began to suspect that he had no idea what my problem was, but I wasn't really sure so I simply endured the strange examination. To signal its end, he pulled out his notebook and said: "Antibiotics. Take one tablet three times a day. It'll get worse before it gets better." Glad that I now had a treatment, I dragged myself back to my hotel room with the prescription in hand.

The pain grew worse and worse—just as the doctor had predicted. The doctor must have known what was wrong with me after all. But, when the pain hadn't subsided after three days, I called him. "Increase the dose to five times a day. It's going to hurt for a while more," he said. After two more days of agony, I finally called the international air ambulance. The Swiss doctor diagnosed appendicitis and operated on me immediately. "Why did you wait so long?" he asked me after the surgery. I replied: "It all happened exactly as the doctor said, so I trusted him."

"Ah, you fell victim to the *it'll-get-worse-before-it-gets-better fallacy*. That Corsican doctor had no idea. Probably just the same type of stand-in you find in all the tourist places in high season."

Let's take another example: A CEO is at his wit's end: Sales are in the toilet, the salespeople are unmotivated, and the marketing campaign sank without a trace. In his desperation, he hires a consultant. For $5,000 a day, this man analyzes the company and comes back with his findings: "Your sales department has no vision, and your brand isn't positioned clearly. It's a tricky situation. I can fix it for you—but not overnight. The measures will require sensitivity, and, most likely, sales will fall further before things improve." The CEO hires the consultant. A year later, sales fall, and the same thing happens the next year. Again and again, the consultant stresses that the company's progress corresponds closely to his prediction. As sales continue their slump in the third year, the CEO fires the consultant.

A mere smoke screen, the *it'll-get-worse-before-it-gets-better fallacy* is a variant of the so-called *confirmation bias*. If the problem continues to worsen, the prediction is confirmed. If the situation improves unexpectedly, the customer is happy, and the expert can attribute it to his prowess. Either way he wins.

Suppose you are president of a country and have no idea how to run it. What do you do? You predict "difficult years" ahead, ask your citizens to "tighten their belts," and then promise to improve the situation only after this "delicate stage" of "cleansing," "purification," and "restructuring." Naturally you leave the duration and severity of the period open.

The best evidence of this strategy's success is the religious zealot who believes that before we can experience heaven on earth, the world must be destroyed. Disasters, floods, fires, death—they are all part of the larger plan and must take place. These believers will view any deterioration of the situation as confirmation of the prophecy and any improvement as a gift from God.

In conclusion: If someone says, "It'll get worse before it gets better," you should hear alarm bells ringing. But beware: Situations do exist where things first dip, then improve. For example, a career change requires time and often incorporates loss of pay. The reorganization of a business also takes time. But in all these cases, we can see relatively quickly if the measures are working. The milestones are clear and verifiable. Look to these rather than to the heavens.

13

Even True Stories Are Fairy Tales
Story Bias

L ife is a muddle, as intricate as a Gordian knot. Imagine an invisible Martian decides to follow you around with an equally invisible notebook, recording what you do, think, and dream. The rundown of your life would consist of entries such as "drank coffee, two sugars," "stepped on a thumbtack and swore like a sailor," "dreamed that I kissed the neighbor," "booked vacation, Maldives, now nearly out of money," "found hair sticking out of ear, plucked it right away," and so on. We like to knit this jumble of details into a neat story. We want our lives to form a pattern that can be easily followed. Many call this guiding principle "meaning." If our story advances evenly over the years, we refer to it as "identity." "We try on stories as we try on clothes," said Max Frisch, a famous Swiss novelist.

We do the same with world history, shaping the details into a consistent story. Suddenly we "understand" certain things, for example, why the Treaty of Versailles led to the Second World War, or why Alan Greenspan's loose monetary policy created the collapse of Lehman Brothers. We comprehend why the

Iron Curtain had to fall or why Harry Potter became a best-seller. Here, we speak about "understanding," but these things cannot be understood in the traditional sense. We simply build the meaning into them afterward. Stories are dubious entities. They simplify and distort reality and filter things that don't fit. But apparently we cannot do without them. Why remains unclear. What is clear is that people first used stories to explain the world, before they began to think scientifically, making mythology older than philosophy. This has led to the *story bias*.

In the media, *story bias* rages like wildfire. For example: A car is driving over a bridge when the structure suddenly collapses. What do we read the next day? We hear the tale of the unlucky driver, where he came from, and where he was going. We read his biography: born somewhere, grew up somewhere else, earned a living as something. If he survives and can give interviews, we hear exactly how it felt when the bridge came crashing down. The absurd thing: Not one of these stories explains the underlying cause of the accident. Skip past the driver's account—and consider the bridge's construction: Where was the weak point? Was it fatigue? If not, was the bridge damaged? If so, by what? Was a proper design even used? Where are there other bridges of the same design? The problem with all these questions is that, though valid, they just don't make for a good yarn. Stories attract us; abstract details repel us. Consequently, entertaining side issues and backstories are prioritized over relevant facts. (On the upside, if it were not for this, we would be stuck with only nonfiction books.)

Here are two stories from the English novelist E. M. Forster. Which one would you remember better? (a) "The king died, and the queen died." (b) "The king died, and the queen died of grief."

Most people will retain the second story more easily. Here, the two deaths don't just take place successively; they are emotionally linked. Story A is a factual report, but story B has "meaning." According to information theory, we should be able to hold on to A better: It is shorter. But our brains don't work that way.

Advertisers have learned to capitalize on this, too. Instead of focusing on an item's benefits, they create a story around it. Objectively speaking, narratives are irrelevant. But still we find them irresistible. Google illustrated this masterfully in its Super Bowl commercial from 2010, "Google Parisian Love." Take a look at it on YouTube.

From our own life stories to global events, we shape everything into meaningful stories. Doing so distorts reality and affects the quality of our decisions, but there is a remedy: Pick these apart. Ask yourself: What are they trying to hide? Visit the library and spend half a day reading old newspapers. You will see that events that today look connected weren't so at the time. To experience the effect once more, try to view your life story out of context. Dig into your old journals and notes, and you'll see that your life has not followed a straight line leading to today, but has been a series of unplanned, unconnected events and experiences, as we will see in the next chapter.

Whenever you hear a story, ask yourself: Who is the sender, what are his intentions, and what did he hide under the rug? The omitted elements might not be of relevance. But, then again, they might be even more relevant than the elements featured in the story, such as when "explaining" a financial crisis or the "cause" of war. The real issue with stories: They give us a false sense of understanding, which inevitably leads us to take bigger risks and urges us to take a stroll on thin ice.

14

Why You Should Keep a Diary

Hindsight Bias

I came across the diaries of my great-uncle recently. In 1932, he emigrated from a tiny Swiss village to Paris to seek his fortune in the movie industry. In August 1940, two months after Paris was occupied, he noted: "Everyone is certain that the Germans will leave by the end of year. Their officers also confirmed this to me. England will fall as fast as France did, and then we will finally have our Parisian lives back—albeit as part of Germany." The occupation lasted four years.

In today's history books, the German occupation of France seems to form part of a clear military strategy. In retrospect, the actual course of the war appears the most likely of all scenarios. Why? Because we have fallen victim to the *hindsight bias*.

Let's take a more recent example: In 2007, economic experts painted a rosy picture for the coming years. However, just twelve months later, the financial markets imploded. Asked about the crisis, the same experts enumerated its causes: monetary expansion under Greenspan, lax validation of mortgages, corrupt rating agencies, low capital requirements, and so forth. In hindsight, the reasons for the crash seem painfully obvious.

The *hindsight bias* is one of the most prevailing fallacies of all. We can aptly describe it as the "I told you so" phenomenon: In retrospect, everything seems clear and inevitable. If a CEO becomes successful due to fortunate circumstances, he will, looking back, rate the probability of his success a lot higher than it actually was. Similarly, following Ronald Reagan's massive election victory over Jimmy Carter in 1980, commentators announced his appointment to be foreseeable, even though the election lay on a knife edge until a few days before the final vote. Today, business journalists opine that Google's dominance was predestined, even though each of them would have snorted had such a prediction been made in 1998. One particularly blundering example: Nowadays it seems tragic, yet completely plausible, that a single shot in Sarajevo in 1914 would totally upturn the world for thirty years and cost fifty million lives. Every child learns this historical detail in school. But back then, nobody would have dreamed of such an escalation. It would have sounded too absurd.

So why is the *hindsight bias* so perilous? Well, it makes us believe we are better predictors than we actually are, causing us to be arrogant about our knowledge and consequently to take too much risk. And not just with global issues: "Have you heard? Sylvia and Chris aren't together anymore. It was always going to go wrong, they were just so different." Or: "They were just so similar." Or: "They spent too much time together." Or even: "They barely saw one another."

Overcoming the *hindsight bias* is not easy. Studies have shown that people who are aware of it fall for it just as much as everyone else. So, I'm very sorry, but you've just wasted your time reading this chapter.

If you're still with me, I have one final tip, this time from personal rather than professional experience: Keep a journal. Write down your predictions—for political changes, your career, your weight, the stock market, and so on. Then, from time to time, compare your notes with actual developments. You will be amazed at what a poor forecaster you are. Don't forget to read history, too—not the retrospective, compacted theories compiled in textbooks, but the diaries, oral histories, and historical documents from the period. If you can't live without news, read newspapers from five, ten, or twenty years ago. This will give you a much better sense of just how unpredictable the world is. Hindsight may provide temporary comfort to those overwhelmed by complexity, but as for providing deeper revelations about how the world works, you'll benefit by looking elsewhere.

15

Why You Systematically Overestimate Your Knowledge and Abilities

Overconfidence Effect

My favorite musician, Johann Sebastian Bach, was anything but a one-hit wonder. He composed numerous works. How many there were I will reveal at the end of this chapter. But for now, here's a small assignment: How many concertos do you think Bach composed? Choose a range, for example, between one hundred and five hundred, aiming for an estimate that is 98 percent correct and only 2 percent off.

How much confidence should we have in our own knowledge? Psychologists Howard Raiffa and Marc Alpert, wondering the same thing, have interviewed hundreds of people in this way. Sometimes they have asked participants to estimate the total egg production in the United States or the number of physicians and surgeons listed in the Yellow Pages of the phone directory for Boston or the number of foreign automobiles imported into the United States, or even the toll collections of the Panama Canal in millions of dollars. Subjects could choose any range they liked, with the aim of being wrong no more than 2 percent of the time. The results were amazing. In the final tally, instead of just 2 percent, they proved incorrect 40 percent of the time. The researchers dubbed this amazing phenomenon the *overconfidence effect*.

The *overconfidence effect* also applies to forecasts, such as stock market performance over a year or your firm's profits over three years. We systematically overestimate our knowledge and our ability to predict—on a massive scale. The *overconfidence effect* does not deal with whether single estimates are correct or not. Rather, it measures the difference between what people really know and what they *think* they know. What's surprising is this: Experts suffer even more from the *overconfidence effect* than laypeople do. If asked to forecast oil prices in five years' time, an economics professor will be as wide of the mark as a zookeeper will. However, the professor will offer his forecast with certitude.

The *overconfidence effect* does not stop at economics: In surveys, 84 percent of Frenchmen estimate that they are above-average lovers. Without the *overconfidence effect*, that figure should be exactly 50 percent—after all, the statistical "median" means 50 percent should rank higher and 50 percent should rank lower. In another survey, 93 percent of the U.S. students estimated to be "above average" drivers. And 68 percent of the faculty at the University of Nebraska rated themselves in the top 25 percent for teaching ability. Entrepreneurs and those wishing to marry also deem themselves to be different: They believe they can beat the odds. In fact, entrepreneurial activity would be a lot lower if the *overconfidence effect* did not exist. For example, every restaurateur hopes to establish the next Michelin-starred restaurant, even though statistics show that most close their doors after just three years. The return on investment in the restaurant business lies chronically below zero.

Hardly any major projects exist that are completed in less time and at a lower cost than forecasted. Some delays and cost overruns are even legendary, such as the Airbus A400M, the Sydney Opera House, and Boston's Big Dig. The list can be added to at will. Why is that? Here, two effects act in unison. First, you have the classic *overconfidence effect*. Second, those with a direct interest in the project have an incentive to underestimate the costs: Consultants, contractors, and suppliers seek follow-up orders. Builders feel bolstered by the optimistic figures, and through their activities, politicians get more votes. We will examine this *strategic misrepresentation* (chapter 89) later in the book.

What makes the *overconfidence effect* so prevalent and its effect so confounding is that it is not driven by incentives; it is raw and innate. And it's not counterbalanced by the opposite effect, "underconfidence," which doesn't exist. No surprise to some readers: The *overconfidence effect* is more pronounced in men—women tend not to overestimate their knowledge and abilities as much. Even more troubling: Optimists are not the only victims of the *overconfidence effect*. Even self-proclaimed pessimists overrate themselves—just less extremely.

In conclusion: Be aware that you tend to overestimate your knowledge. Be skeptical of predictions, especially if they come from so-called experts. And with all plans, favor the pessimistic scenario. This way, you have a chance of judging the situation somewhat realistically.

Back to the question from the beginning: Johann Sebastian Bach composed 1,127 works that survived to this day. He may have composed considerably more, but they are lost.

Don't Take News Anchors Seriously
Chauffeur Knowledge

After receiving the Nobel Prize in Physics in 1918, Max Planck went on tour across Germany. Wherever he was invited, he delivered the same lecture on new quantum mechanics. Over time, his chauffeur grew to know it by heart: "It has to be boring giving the same speech each time, Professor Planck. How about I do it for you in Munich? You can sit in the front row and wear my chauffeur's cap. That'd give us both a bit of variety." Planck liked the idea, so that evening the driver held a long lecture on quantum mechanics in front of a distinguished audience. Later, a physics professor stood up with a question. The driver recoiled: "Never would I have thought that someone from such an advanced city as Munich would ask such a simple question! My chauffeur will answer it."

According to Charlie Munger, one of the world's best investors (and from whom I have borrowed this story), there are two types of knowledge. First, we have *real* knowledge. We see it in people who have committed a large amount of time and effort to understanding a topic. The second type is *chauffeur knowledge*—knowledge from people who have learned to put on a show. Maybe they have a great voice or good hair, but the knowledge they espouse is not their own. They reel off eloquent words as if reading from a script.

Unfortunately, it is increasingly difficult to separate true knowledge from *chauffeur knowledge*. With news anchors, however, it is still easy. These are actors. Period. Everyone knows it. And yet it continues to astound me how much respect these perfectly coiffed script readers enjoy, not to mention how much they earn, moderating panels about topics they barely fathom.

With journalists, it is more difficult. Some have acquired true knowledge. Often they are veteran reporters who have specialized for years in a clearly defined area. They make a serious effort to understand the complexity of a subject and to communicate it. They tend to write long articles that highlight a variety of cases and exceptions. The majority of journalists, however, fall into the category of chauffeur. They conjure up articles off the tops of their heads or, rather, from Google searches. Their texts are one-sided, short, and—often as compensation for their patchy knowledge—snarky and self-satisfied in tone.

The same superficiality is present in business. The larger a company, the more the CEO is expected to possess "star quality." Dedication, solemnity, and reliability are undervalued, at least at the top. Too often shareholders and business journalists seem to believe that showmanship will deliver better results, which is obviously not the case.

To guard against the chauffeur effect, Warren Buffett, Munger's business partner, has coined a wonderful phrase, the "circle of competence": What lies inside this circle you understand intuitively; what lies outside, you may only partially comprehend. One of Munger's best pieces of advice is: "You have to stick within what I call your circle of competence. You have to know what you understand and what you don't understand. It's not terribly important how big the circle is. But it is terribly important that you know where the perimeter is." Munger underscores this: "So you have to figure out what your own aptitudes are. If you play games where other people have the aptitudes and you don't, you're going to lose. And that's as close to certain as any prediction that you can make. You have to figure out where you've got an edge. And you've got to play within your own circle of competence."

In conclusion: Be on the lookout for *chauffeur knowledge*. Do not confuse the company spokesperson, the ringmaster, the newscaster, the schmoozer, the verbiage vendor, or the cliché generator with those who possess true knowledge. How do you recognize the difference? There is a clear indicator: True experts recognize the limits of what they know and what they do *not* know. If they find themselves outside their circle of competence, they keep quiet or simply say, "I don't know." This they utter unapologetically, even with a certain pride. From chauffeurs, we hear every line except this.

17

You Control Less Than You Think

Illusion of Control

Every day, shortly before nine o'clock, a man with a red hat stands in a square and begins to wave his cap around wildly. After five minutes, he disappears. One day, a policeman comes up to him and asks: "What are you doing?" "I'm keeping the giraffes away." "But there aren't any giraffes here." "Well, I must be doing a good job, then."

A friend with a broken leg was stuck in bed and asked me to pick up a lottery ticket for him. I went to the store, checked a few boxes, wrote his name on it, and paid. As I handed him the copy of the ticket, he balked: "Why did *you* fill it out? I wanted to do that. I'm never going to win anything with your numbers!" "Do you really think it affects the draw if *you* pick the numbers?" I inquired. He looked at me blankly.

In casinos, most people throw the dice as hard as they can if they need a high number and as gingerly as possible if they are hoping for a low number—which is as nonsensical as football fans thinking they can swing a game by gesticulating in front of the TV. Unfortunately they share this illusion with many people who also seek to influence the world by sending out the "right" thoughts (i.e., vibrations, positive energy, karma . . .).

The *illusion of control* is the tendency to believe that we can influence something over which we have absolutely no sway. This was discovered in 1965 by two researchers, Jenkins and Ward. Their experiment was simple, consisting of just two switches and a light. The men were able to adjust when the switches connected to the light and when not. Even when the light flashed on and off at random, subjects were still convinced that they could influence it by flicking the switches.

Or consider this example: An American researcher has been investigating acoustic sensitivity to pain. For this, he placed people in sound booths and increased the volume until the subjects signaled him to stop. The two rooms, A and B, were identical, save one thing: Room B had a red panic button on the wall. The button was purely for show, but it gave participants the feeling that they were in control of the situation, leading them to withstand significantly more noise. If you have read Aleksandr Solzhenitsyn, Primo Levi, or Viktor Frankl, this finding will not surprise you: The idea that people can influence their destiny, even by a fraction, encouraged these prisoners not to give up hope.

Crossing the street in Los Angeles is a tricky business, but luckily, at the press of a button, we can stop traffic. Or can we? The button's real purpose is to make us believe we have an influence on the traffic lights, and thus we're better able to endure the wait for the signal to change with more patience. The same goes for "door-open" and "door-close" buttons in elevators: Many are not even connected to the electrical panel. Such tricks are also designed in open-plan offices: For some people it will always be too hot, for others, too cold. Clever technicians create the *illusion of control* by installing fake temperature dials. This reduces energy bills—and complaints. Such ploys are called "placebo buttons" and they are being pushed in all sorts of realms.

Central bankers and government officials employ placebo buttons masterfully. Take, for instance, the federal funds rate, which is an extreme short-term rate—an overnight rate, to be precise. While this rate doesn't affect long-term interest rates (which are a function of supply and demand, and which are an important factor in investment decisions), the stock market, nevertheless, reacts frenetically to its every change. Nobody understands why overnight interest rates can have such an effect on the market, but everybody thinks they do, and so they do. The same goes for pronouncements made by the chairman of the Federal Reserve; markets move, even though these statements inject little of tangible value into the real economy. They are merely sound waves. And still we allow economic heads to continue to play with the illusory dials. It would be a real wake-up call if all involved realized the truth—that the world economy is a fundamentally uncontrollable system.

And you? Do you have everything under control? Probably less than you think. Do not think you command your way through life like a Roman emperor. Rather, you are the man with the red hat. Therefore, focus on the few things of importance that you can really influence. For everything else: *Que sera, sera.*

18

Never Pay Your Lawyer by the Hour
Incentive Super-Response Tendency

To control a rat infestation, French colonial rulers in Hanoi in the nineteenth century passed a law: For every dead rat handed in to the authorities, the catcher would receive a reward. Yes, many rats were destroyed, but many were also bred specially for this purpose.

In 1947, when the Dead Sea Scrolls were discovered, archaeologists set a finder's fee for each new parchment. Instead of lots of extra scrolls being found, they were simply torn apart to increase the reward. Similarly, in China in the nineteenth century, an incentive was offered for finding dinosaur bones. Farmers located a few on their land, broke them into pieces, and cashed in. Modern incentives are no better: Company boards promise bonuses for achieved targets. And what happens? Managers invest more energy in trying to lower the targets than in growing the business.

These are examples of the *incentive super-response tendency*. Credited to Charlie Munger, this titanic name describes a rather trivial observation: People respond to incentives by doing what is in their best interests. What is noteworthy is, first, how quickly and radically people's behavior changes when incentives come into play or are altered, and second, the fact that people respond to the incentives themselves, and not the grander intentions behind them.

Good incentive systems comprise both intent and reward. An example: In ancient Rome, engineers were made to stand *underneath* the construction at their bridges' opening ceremonies. Poor incentive systems, on the other hand, overlook and sometimes even pervert the underlying aim. For example, censoring a book makes its contents more famous, and rewarding bank employees for each loan sold leads to a miserable credit portfolio. Making CEO pay public didn't dampen the astronomical salaries; to the contrary, it pushed them upward. Nobody wants to be the loser CEO in his industry.

Do you want to influence the behavior of people or organizations? You could always preach about values and visions or you could appeal to reason. But in nearly every case, incentives work better. These need not be monetary; anything is possible, from good grades to Nobel Prizes to special treatment in the afterlife.

For a long time I tried to understand what made well-educated nobles from the Middle Ages bid adieu to their comfortable lives, swing themselves up onto horses, and take part in the Crusades. They were well aware that the arduous ride to Jerusalem lasted at least six months and passed directly through enemy territory; yet they took the risk. And then it came to me: The answer lies in incentive systems. If they came back alive, they could keep the spoils of war and live out their days as rich men. If they died, they automatically passed on to the afterlife as martyrs—with all the benefits that came with it. It was win-win.

Imagine for a moment that, instead of demanding enemies' riches, warriors and soldiers charged by the hour. We would effectively be incentivizing them to take as long as possible, right? So why do we do just this with lawyers, architects, consultants, accountants, and driving instructors? My advice: Forget hourly rates and always negotiate a fixed price in advance.

Be wary, too, of investment advisers endorsing particular financial products. They are not interested in your financial well-being, but in earning a commission on these products. The same goes for entrepreneurs' and investment bankers' business plans. These are often worthless because, again, the vendors have their own interests at heart. What is the old adage? "Never ask a barber if you need a haircut."

In conclusion: Keep an eye out for the *incentive super-response tendency*. If a person's or an organization's behavior confounds you, ask yourself what incentive might lie behind it. I guarantee you that you'll be able to explain 90 percent of the cases this way. What makes up the remaining 10 percent? Passion, idiocy, psychosis, or malice.

19

The Dubious Efficacy of Doctors, Consultants, and Psychotherapists

Regression to Mean

His back pain was sometimes better, sometimes worse. There were days when he felt like he could move mountains, and those when he could barely move. If that was the case—fortunately it happened only rarely—his wife would drive him to the chiropractor. The next day he felt much more mobile and recommended the therapist to everyone.

Another man, younger and with a respectable golf handicap of 12, gushed in a similar fashion about his golf instructor. Whenever he played miserably, he booked an hour with the pro, and, lo and behold, in the next game he fared much better.

A third man, an investment adviser at a major bank, invented a sort of "rain dance" that he performed in the restroom every time his stocks had performed extremely badly. As absurd as it seemed, he felt compelled to do it: Things always improved afterward.

What links the three men is a fallacy: the *regression-to-mean* delusion.

Suppose your region is experiencing a record period of cold weather. In all probability, the temperature will rise in the next few days—back toward the monthly average. The same goes for extreme heat, drought, or rain. Weather fluctuates around a mean. The same is true for chronic pain, golf handicaps, stock market performance, luck in love, subjective happiness, and test scores. In short, the crippling back pain would most likely have improved without a chiropractor. The handicap would have returned to 12 without additional lessons. And the performance of the investment adviser would also have shifted back toward the market average—with or without the restroom dance.

Extreme performances are interspersed with less extreme ones. The most successful stock picks from the past three years are hardly going to be the most successful stocks in the coming three years. Knowing this, you can appreciate why some athletes would rather not make it on to the front pages of the newspapers: Subconsciously they know that the next time they race, they probably won't achieve the same top result—which has nothing to do with the media attention, but with natural variations in performance.

Or take the example of a division manager who wants to improve employee morale by sending the least motivated 3 percent of the workforce on a course. The result? The next time he looks at motivation levels, the same people will not make up the bottom few—there will be others. Was the course worth it? Hard to say, since the group's motivation levels would probably have returned to their personal norms even without the training. The situation is similar with patients who are hospitalized for depression. They usually leave the clinic feeling a little better. It is quite possible, however, that the stay contributed absolutely nothing.

Another example: In Boston, the lowest-performing schools were entered into a complex support program. The following year, the schools had moved up in the rankings, an improvement that the authorities attributed to the program rather than to natural *regression to mean*.

Ignoring *regression to mean* can have destructive consequences, such as teachers (or managers) concluding that the stick is better than the carrot. For example, following a test, the highest-performing students are praised and the lowest are castigated. In the next exam, other students will probably—purely coincidentally—achieve the highest and lowest scores. Thus, the teacher concludes that reproach helps and praise hinders: a fallacy that keeps on giving.

In conclusion: When you hear stories such as: "I was sick, went to the doctor, and got better a few days later" or "the company had a bad year, so we got a consultant in, and now the results are back to normal," look out for our old friend, the *regression-to-mean* error.

20

Never Judge a Decision by Its Outcome

Outcome Bias

A quick hypothesis: Say one million monkeys speculate on the stock market. They buy and sell stocks like crazy and, of course, completely at random. What happens? After one week, about half of the monkeys will have made a profit and the other half a loss. The ones that made a profit can stay; the ones that made a loss you send home. In the second week, one half of the monkeys will still be riding high, while the other half will have made a loss and are sent home. And so on. After ten weeks, about one thousand monkeys will be left—those who have always invested their money well. After twenty weeks, just one monkey will remain—this one always, without fail, chose the right stocks and is now a billionaire. Let's call him the success monkey.

How does the media react? It will pounce on this animal to understand its "success principles." And they will find some: Perhaps the monkey eats more bananas than the others. Perhaps he sits in another corner of the cage. Or maybe he swings headlong through the branches, or he takes long, reflective pauses while grooming. He must have some recipe for success, right? How else could he perform so brilliantly? Spot-on for two years—and that from a simple monkey? Impossible!

The monkey story illustrates the *outcome bias*: We tend to evaluate decisions based on the result rather than on the decision process. This fallacy is also known as the "historian error." A classic example is the Japanese attack on Pearl Harbor. Should the military base have been evacuated or not? From today's perspective: obviously, for there was plenty of evidence that an attack was imminent. However, only in retrospect do the signals appear so clear. At the time, in 1941, there was a plethora of contradictory signals. Some pointed to an attack; others did not. To assess the quality of the decision, we must use the information available at the time, filtering out everything we know about it postattack (particularly that it did indeed take place).

Another experiment: You must evaluate the performance of three heart surgeons. To do this, you ask each to carry out a difficult operation five times. Over the years, the probability of dying from these procedures has stabilized at 20 percent. With surgeon A, no one dies. With surgeon B, one patient dies. With surgeon C, two die. How do you rate the performances of A, B, and C? If you think like most people, you rate A the best, B the second best, and C the worst. And thus you've just fallen for the *outcome bias*. You can guess why: The samples are too small, rendering the results meaningless. You can only really judge a surgeon if you know something about the field, and then carefully monitor the preparation and execution of the operation. In other words, you assess the process and not the result. Alternatively, you could employ a larger sample: one hundred or one thousand operations if you have enough patients who need this particular operation. For now it is enough to know that, with an average surgeon, there is a 33 percent chance that no one will die, a 41 percent chance that one person will die, and a 20 percent chance that two people will die. That's a simple probability calculation. What stands out: There is no huge difference between zero dead and two dead. To assess the three surgeons purely on the basis of the outcomes would be not only negligent, but also unethical.

In conclusion: Never judge a decision purely by its result, especially when randomness and "external factors" play a role. A bad result does not automatically indicate a bad decision and vice versa. So rather than tearing your hair out about a wrong decision, or applauding yourself for one that may have only coincidentally led to success, remember why you chose what you did. Were your reasons rational and understandable? Then you would do well to stick with that method, even if you didn't strike it lucky last time.

21

Less Is More

Paradox of Choice

My sister and her husband bought an unfinished house a little while ago. Since then, we haven't been able to talk about anything else. The sole topic of conversation for the past two months has been bathroom tiles: ceramic, granite, marble, metal, stone, wood, glass, and every type of laminate known to man. Rarely have I seen my sister in such anguish. "There are just too many to choose from," she exclaims, throwing her hands in the air and returning to the tile catalog, her constant companion.

I've counted and researched: My local grocery store stocks 48 varieties of yogurt, 134 types of red wine, 64 different cleaning products, and a grand total of 30,000 items. Amazon, the Internet bookseller, has two million titles available. Nowadays, people are bombarded with options, such as hundreds of mental disorders, thousands of different careers, even more holiday destinations, and an infinite variety of lifestyles. There has never been more choice.

When I was young, we had three types of yogurt, three television channels, two churches, two kinds of cheese (mild or strong), one type of fish (trout), and one telephone provided by the Swiss Post. The black box with the dial served no other purpose than making calls, and that did us just fine. In contrast, anyone who enters a cell-phone store today runs the risk of being flattened by an avalanche of brands, models, and contract options.

And yet selection is the yardstick of progress. It is what sets us apart from planned economies and the Stone Age. Yes, abundance makes you giddy, but there is a limit. When it is exceeded, a surfeit of choices destroys quality of life. The technical term for this is the *paradox of choice.*

In his book of the same title, psychologist Barry Schwartz describes why this is so. First, a large selection leads to inner paralysis. To test this, a supermarket set up a stand where customers could sample twenty-four varieties of jelly. They could try as many as they liked and then buy them at a discount. The next day, the owners carried out the same experiment with only six flavors. The result? They sold ten times more jelly on day two. Why? With such a wide range, customers could not come to a decision, so they bought nothing. The experiment was repeated several times with different products. The results were always the same.

Second, a broader selection leads to poorer decisions. If you ask young people what is important in a life partner, they reel off all the usual qualities: intelligence, good manners, warmth, the ability to listen, a sense of humor, and physical attractiveness. But do they actually take these criteria into account when choosing someone? In the past, a young man from a village of average size could choose among maybe twenty girls of similar age with whom he went to school. He knew their families and vice versa, leading to a decision based on several well-known attributes. Nowadays, in the era of online dating, millions of potential partners are at our disposal. It has been proven that the stress caused by this mind-boggling variety is so large that the male brain reduces the decision to one single criterion: physical attractiveness. The consequences of this selection process you already know—perhaps even from personal experience.

Finally, large selection leads to discontent. How can you be sure you are making the right choice when two hundred options surround and confound you? The answer is: You cannot. The more choice you have, the more unsure and therefore dissatisfied you are afterward.

So what can you do? Think carefully about what you want before you inspect existing offers. Write down these criteria and stick to them rigidly. Also, realize that you can never make a perfect decision. Aiming for this is, given the flood of possibilities, a form of irrational perfectionism. Instead, learn to love a "good" choice. Yes, even in terms of life partners. Only the best will do? In this age of unlimited variety, rather the opposite is true: "Good enough" is the new optimum (except, of course, for you and me).

22

You Like Me, You Really, Really Like Me

Liking Bias

Kevin has just bought two boxes of fine Margaux. He rarely drinks wine—not even Bordeaux—but the sales assistant was so nice, not fake or pushy, just really likable. So he bought them.

Joe Girard is considered the most successful car salesman in the world. His tip for success: "There's nothing more effective in selling anything than getting the customer to believe, really believe, that you like him and care about him." Girard doesn't just talk the talk: His secret weapon is sending a card to his customers each month. Just one sentence salutes them: "I like you."

The *liking bias* is startlingly simple to understand and yet we continually fall prey to it. It means this: The more we like someone, the more inclined we are to buy from or help that person. Still, the question remains: What does "likable" even mean? According to research, we see people as pleasant, if (a) they are outwardly attractive, (b) they are similar to us in terms of origin, personality, or interests, and (c) they like us. Consequently, advertising is full of attractive people. Ugly people seem unfriendly and don't even make it into the background (see A). In addition to engaging super-attractive types, advertising also employs "people like you and me" (see B)—those who are similar in appearance, accent, or background. In short, the more similar, the better. Mirroring is a standard technique in sales to get exactly this effect. Here, the salesperson tries to copy the gestures, language, and facial expressions of his prospective client. If the buyer speaks very slowly and quietly, often scratching his head, it makes sense for the seller to speak slowly and quietly, and to scratch his head now and then, too. That makes him likable in the eyes of the buyer, and thus a business deal is more likely. Finally, it's not unheard of for advertisers to pay us compliments: How many times have you bought something "because you're worth it"? Here factor C comes into play: We find people appealing if they like us. Compliments work wonders, even if they ring hollow as a drum.

So-called multilevel marketing (selling through personal networks) works solely because of the *liking bias*. Though there are excellent plastic containers in the supermarket for a quarter of the price, Tupperware generates annual revenues of $2 billion. Why? The friends who hold the Tupperware parties meet the second and third congeniality standard perfectly.

Aid agencies employ the *liking bias* to great effect. Their campaigns use beaming children or women almost exclusively. Never will you see a stone-faced, wounded guerrilla fighter staring at you from billboards—even though he also needs your support. Conservation organizations also carefully select who gets the starring role in their advertisements. Have you ever seen a World Wildlife Fund brochure filled with spiders, worms, algae, or bacteria? They are perhaps just as endangered

as pandas, gorillas, koalas, and seals—and even more important for the ecosystem. But we feel nothing for them. The more
human a creature acts, the more similar it is to us, the more we
like it. The bone skipper fly is extinct? Too bad.

Politicians, too, are maestros of the *liking bias*. Depending
on the makeup and interests of an audience, they emphasize
different topics, such as residential area, social background, or
economic issues. And they flatter us: Each potential voter is
made to feel like an indispensable member of the team: "Your
vote counts!" Of course your vote counts, but only by the tiniest
of fractions, bordering on the irrelevant.

A friend who deals in oil pumps told me how he once closed
an eight-figure deal for a pipeline in Russia. "Bribery?" I inquired. He shook his head. "We were chatting, and suddenly
we got on to the topic of sailing. It turned out that both of us—
the buyer and me—were die-hard 470 dinghy fans. From that
moment on, he liked me; I was a friend. So the deal was sealed.
Amiability works better than bribery."

So, if you are a salesperson, make buyers think you like
them, even if this means outright flattery. And if you are a consumer, always judge a product independent of who is selling it.
Banish the salespeople from your mind or, rather, pretend you
don't like them.

23

Don't Cling to Things
Endowment Effect

The BMW gleamed in the parking lot of the used-car dealership. Although it had a few miles on the odometer, it looked in perfect condition. I know a little about used cars, and to me, it was worth around $40,000. However, the salesman was pushing for $50,000 and wouldn't budge a dime. When he called the next week to say he would accept $40,000 after all, I went for it. The next day, I took it out for a spin and stopped at a gas station. The owner came out to admire the car—and proceeded to offer me $53,000 in cash on the spot. I politely declined. Only on the way home did I realize how ridiculous I was to have said no. Something that I considered worth $40,000 had passed into my possession and suddenly taken on a value of more than $53,000. If I were thinking purely rationally, I would have sold the car immediately. But, alas, I'd fallen under the influence of the *endowment effect*. We consider things to be more valuable the moment we own them. In other words, if we are selling something, we charge more for it than what we ourselves would be willing to spend.

To probe this, psychologist Dan Ariely conducted the following experiment: In one of his classes, he raffled tickets to a major basketball game, then polled the students to see how much they thought the tickets were worth. The empty-handed students estimated around $170, whereas the winning students would not sell it below an average of $2,400. The simple fact of ownership makes us add zeros to the selling price.

In real estate, the *endowment effect* is palpable. Sellers become emotionally attached to their houses and thus systematically overestimate their value. They balk at the market price, expecting buyers to pay more—which is completely absurd since this excess is little more than sentimental value.

Richard Thaler performed an interesting classroom experiment at Cornell University to measure the *endowment effect*. He distributed coffee mugs to half of the students and told them they could either take the mug home or sell it at a price they could specify. The other half of the students who didn't get a mug were asked how much they would be willing to pay for a mug. In other words, Thaler set up a market for coffee mugs. One would expect that roughly 50 percent of the students would be willing to trade—to either sell or buy a mug. But the result was much lower than that. Why? Because the average owner would not sell below $5.25, and the average buyer would not pay more than $2.25 for a mug.

We can safely say that we are better at collecting things than at casting them off. Not only does this explain why we fill our homes with junk, but also why lovers of stamps, watches, and pieces of art part with them so seldomly.

Amazingly, the *endowment effect* affects not only possession, but also near ownership. Auction houses like Christie's and

Sotheby's thrive on this. A person who bids until the end of an auction gets the feeling that the object is practically theirs, thus increasing its value. The would-be owner is suddenly willing to pay much more than planned, and any withdrawal from the bidding is perceived as a loss—which defies all logic. In large auctions, such as those for mining rights or mobile radio frequencies, we often observe the *winner's curse*: Here, the successful bidder turns out to be the economic loser when he gets caught up in the fervor and overbids. I'll offer more insight on the *winner's curse* in chapter 35.

There's a similar effect in the job market. If you are applying for a job and don't get a call back, you have every reason to be disappointed. However, if you make it to the final stages of the selection process and then receive the rejection, the disappointment can be much bigger—irrationally. Either you get the job or you don't; nothing else should matter.

In conclusion: Don't cling to things. Consider your property something that the "universe" (whatever you believe this to be) has bestowed to you temporarily. Keep in mind that it can recoup this (or more) in the blink of an eye.

24

The Inevitability of Unlikely Events

Coincidence

At 7:15 p.m. on March 1, 1950, the fifteen members of the church choir in Beatrice, Nebraska, were scheduled to meet for rehearsal. For various reasons, they were all running late. The minister's family was delayed because his wife still had to iron their daughter's dress. One couple was held back when their car wouldn't start. The pianist wanted to be there thirty minutes early, but he fell into a deep sleep after dinner. And so on. At 7:25 p.m., the church exploded. The blast was heard all around the village. It blew out the walls and sent the roof crashing to the ground. Miraculously, nobody was killed. The fire chief traced the explosion back to a gas leak, even though members of the choir were convinced they had received a sign from God. Hand of God or *coincidence*?

Something last week made me think of my old school friend, Andy, whom I hadn't spoken to in a long time. Suddenly the phone rang. I picked it up, and, lo and behold, it was Andy. "I must be telepathic!" I exclaimed excitedly. But telepathy or *coincidence*?

On October 5, 1990, the *San Francisco Examiner* reported

that Intel would take its rival, AMD, to court. Intel found out that the company was planning to launch a computer chip named AM386, a term that clearly referred to Intel's 386 chip. How Intel came upon the information is remarkable: By pure coincidence, both companies had hired someone named Mike Webb. Both men were staying in the same hotel in California and checked out on the same day. After they had left, the hotel accepted a package for Mike Webb at reception. It contained confidential documents about the AM386 chip, and the hotel mistakenly sent it to Mike Webb of Intel, who promptly forwarded the contents to the legal department.

How likely are stories like that? The Swiss psychiatrist C. G. Jung saw in them the work of an unknown force, which he called synchronicity. But how should a rationally minded thinker approach these accounts? Preferably with a piece of paper and a pencil. Consider the first case, the explosion of the church. Draw four boxes to represent each of the potential events. The first possibility is what actually took place: "choir delayed and church exploded." But there are three other options: "choir delayed and church did not explode," "choir on time and church exploded," and "choir on time and church did not explode." Estimate the frequencies of these events and write them in the corresponding box. Pay special attention to how often the last case has happened: Every day, millions of choirs gather for scheduled rehearsals and their churches don't blow up. Suddenly, the story has lost its unimaginable quality. For all these millions of churches, it would be improbable if something like what happened in Beatrice, Nebraska, didn't take place at least once a century. So, no hand of God. (And anyway, why would God want to blow a church to smithereens?)

Let's apply the same thinking to the phone call. Keep in mind the many occasions when "Andy" thinks of you but doesn't call; when you think of him and he doesn't call; when you don't think of him and he calls; when he doesn't think of you and you call. . . . There is an almost infinite number of occasions when you don't think of him and he doesn't call. But since people spend about 90 percent of their time thinking about others, it is not unlikely that, eventually, two people will think of each other and one of them will pick up the phone. And it must not be just Andy: If you have a hundred other friends, the probability of this happening increases manifold.

We tend to stumble when estimating probabilities. If someone says "never," I usually register this as a minuscule probability greater than zero since "never" cannot be compensated by a negative probability.

In sum: Let's not get too excited. Improbable coincidences are precisely that: rare but very possible events. It's not surprising when they finally happen. What would be more surprising is if they never came to be.

25

The Calamity of Conformity

Groupthink

Have you ever bitten your tongue in a meeting? Surely. You sit there, say nothing, and nod along to proposals. After all, you don't want to be the (eternal) naysayer. Moreover, you might not be 100 percent sure why you disagree, whereas the others are unanimous—and far from stupid. So you keep your mouth shut for another day. When everyone thinks and acts like this, *groupthink* is at work: This is where a group of smart people makes reckless decisions because everyone aligns their opinions with the supposed consensus. Thus, motions are passed that each individual group member would have rejected if no peer pressure had been involved. *Groupthink* is a special branch of *social proof*, a flaw that we discussed in chapter 4.

In March 1960, the U.S. Secret Service began to mobilize anticommunist exiles from Cuba, most of them living in Miami, to use against Fidel Castro's regime. In January 1961, two days after taking office, President Kennedy was informed about the secret plan to invade Cuba. Three months later, a key meeting took place at the White House, where Kennedy and his advisers all voted in favor of the invasion. On April 17, 1961, a brigade of 1,400 exiled Cubans landed at the Bay of Pigs, on Cuba's south coast, with the help of the U.S. Navy, the Air Force, and the CIA. The aim was to overthrow Castro's government. However, nothing went as planned. On the first day, not a single supply ship reached the coast. The Cuban air force sank the first two, and the next two turned around and fled back to the United States. A day later, Castro's army completely surrounded the brigade. On the third day, the 1,200 survivors were taken into custody and sent to military prisons.

Kennedy's invasion of the Bay of Pigs is regarded as one of the biggest flops in American foreign policy. That such an absurd plan was ever agreed upon, never mind put into action, is astounding. All of the assumptions that spoke in favor of the invasion were erroneous. For example, Kennedy's team completely underestimated the strength of Cuba's air force. Also, it was expected that, in an emergency, the brigade would be able to hide in the Escambray Mountains and carry out an underground war against Castro from there. A glance at the map shows that the refuge was 100 miles away from the Bay of Pigs, with an insurmountable swamp in between. And yet Kennedy and his advisers were among the most intelligent people to ever run an American government. What went wrong between January and April 1961?

Psychology professor Irving Janis has studied many fiascoes. He concluded that they share the following pattern: Members of a close-knit group cultivate team spirit by (unconsciously) building illusions. One of these fantasies is a belief in invincibility: "If both our leader [in this case, Kennedy] and the group are confident that the plan will work, then luck will be on our side." Next comes the illusion of unanimity: If the others are of the same opinion, any dissenting view must be wrong. No one wants to be the naysayer that destroys team unity. Finally, each person is happy to be part of the group. Expressing reservations could mean exclusion from it. In our evolutionary past, such banishment guaranteed death; hence our strong urge to remain in the group's favor.

Groupthink is no stranger in the business world. A classic example is the fate of the world-class airline Swissair. Here, a group of highly paid consultants rallied around the former CEO and, bolstered by the euphoria of past successes, they developed a high-risk expansion strategy (including the acquisition of several European airlines). The zealous team built up such a strong consensus that even rational reservations were suppressed, leading to the airline's collapse in 2001.

If you ever find yourself in a tight, unanimous group, you must speak your mind, even if your team does not like it. Question tacit assumptions, even if you risk expulsion from the warm nest. And, if you lead a group, appoint someone as devil's advocate. She will not be the most popular member of the team, but she might be the most important.

26

Why You'll Soon Be Playing Mega Trillions

Neglect of Probability

Two games of chance: In the first, you can win $10 million, and in the second, $10,000. Which do you play? If you win the first game, it changes your life completely: You can quit your job, tell your boss where to go, and live off the winnings. If you hit the jackpot in the second game, you can take a nice vacation in the Caribbean, but you'll be back at your desk quick enough to see your postcard arrive. The probability of winning is one in 100 million in the first game, and one in 10,000 in the second game. So which do you choose?

Our emotions draw us to the first game, even though the second is ten times better, objectively considered (expected win times probability). Therefore, the trend is toward ever-larger jackpots—Mega Millions, Mega Billions, Mega Trillions—no matter how small the odds are.

In a classic experiment from 1972, participants were divided into two groups. The members of the first group were told that they would receive a small electric shock. In the second group, subjects were told that the risk of this happening was only 50 percent. The researchers measured physical anxiety (heart rate, nervousness, sweating, etc.) shortly before commencing. The result were, well, shocking: There was absolutely no difference. Participants in both groups were equally stressed. Next, the researchers announced a series of reductions in the probability of a shock for the second group: from 50 percent to 20 percent, then 10 percent, then 5 percent. The result: still no difference! However, when they declared they would increase the *strength* of the expected current, both groups' anxiety levels rose—again, by the same degree. This illustrates that we respond to the expected *magnitude* of an event (the size of the jackpot or the amount of electricity), but not to its *likelihood*. In other words: We lack an intuitive grasp of probability.

The proper term for this is *neglect of probability*, and it leads to errors in decision making. We invest in start-ups because the potential profit makes dollar signs flash before our eyes, but we forget (or are too lazy) to investigate the slim chances of new businesses actually achieving such growth. Similarly, following extensive media coverage of a plane crash, we cancel flights without really considering the minuscule probability of crashing (which, of course, remains the same before and after such a disaster). Many amateur investors compare their investments solely on the basis of yield. For them, Google shares with a return of 20 percent must be twice as good as property that returns 10 percent. That's wrong. It would be a lot smarter to also consider both investments' risks. But then again, we have no natural feel for this, so we often turn a blind eye to it.

Back to the experiment with the electric shocks: In group B, the probability of getting a jolt was further reduced: from 5 percent to 4 percent to 3 percent. Only when the probability reached zero did group B respond differently than group A. To us, 0 percent risk seems infinitely better than a (highly improbable) 1 percent risk.

To test this, let's examine two methods of treating drinking water. Suppose a river has two equally large tributaries. One is treated using method A, which reduces the risk of dying from contaminated water from 5 percent to 2 percent. The other is treated using method B, which reduces the risk from 1 percent to 0 percent, that is, the threat is completely eliminated. So, method A or B? If you think like most people, you will opt for method B—which is silly because with measure A, 3 percent fewer people die, and with B, just 1 percent fewer. Method A is three times as good! This fallacy is called the "zero-risk bias."

A classic example of this is the U.S. Food Act of 1958, which prohibits food that contains cancer-causing substances. Instituted to achieve zero risk of cancer, this ban sounds good at first, but it ended up leading to the use of more dangerous (but noncarcinogenic) food additives. It is also absurd: As Paracelsus illustrated in the sixteenth century, poisoning is always a question of dosage. Furthermore, this law can never be enforced properly since it is impossible to remove the last "banned" molecule from food. Each farm would have to function like a hyper-sterile computer-chip factory, and the cost of food would increase a hundredfold. Economically, zero risk rarely makes sense. One exception is when the consequences are colossal, such as a deadly, highly contagious virus escaping from a biotech laboratory.

We have no intuitive grasp of risk and thus distinguish poorly among different threats. The more serious the threat and the more emotional the topic (such as radioactivity), the less reassuring a reduction in risk seems to us. Two researchers at the University of Chicago have shown that people are equally afraid of a 99 percent chance as they are of a 1 percent chance of contamination by toxic chemicals. An irrational response, but a common one.

27

Why the Last Cookie in the Jar Makes Your Mouth Water

Scarcity Error

Coffee at a friend's house. We sat trying to make conversation while her three children grappled with one another on the floor. Suddenly I remembered that I had brought some glass marbles with me—a whole bag full. I spilled them out on the floor, in the hope that the little angels would play with them in peace. Far from it: A heated argument ensued. I didn't understand what was happening until I looked more closely. Apparently, among the countless marbles, there was just one blue one, and the children scrambled for it. All the marbles were exactly the same size and shiny and bright. But the blue one had an advantage over the others—it was one of a kind. I had to laugh at how childish children are!

In August 2005, when I heard that Google would launch its own e-mail service, I was dead-set on getting an account. (In the end I did.) At the time, new accounts were very restricted and were given out only by invitation. This made me want one even more. But why? Certainly not because I needed another e-mail account (back then, I already had four), or because Gmail was better than the competition, but simply because not everyone had access to it. Looking back, I have to laugh at how childish adults are!

Rara sunt cara, said the Romans. Rare is valuable. In fact, the *scarcity error* is as old as mankind. My friend with the three children is a part-time real estate agent. Whenever she has an interested buyer who cannot decide, she calls and says: "A doctor from London saw the plot of land yesterday. He liked it a lot. What about you? Are you still interested?" The doctor from London—sometimes it's a professor or a banker—is, of course, fictitious. The effect is very real, though: It causes prospects to see the opportunity disappearing before their eyes, so they act and close the deal. Why? This is the potential shortage of supply, yet again. Objectively, this situation is incomprehensible: Either the prospect wants the land for the set price or he does not—regardless of any doctors from London.

To assess the quality of cookies, Professor Stephen Worchel split participants into two groups. The first group received an entire box of cookies, and the second group just two. In the end, the subjects with just two cookies rated the quality much higher than the first group did. The experiment was repeated several times and always showed the same result.

"Only while stocks last," the ads alert. "Today only," warn the posters. Gallery owners take advantage of the *scarcity error* by placing red "sold" dots under most of their paintings, transforming the remaining few works into rare items that must be snatched up quickly. We collect stamps, coins, vintage cars even when they serve no practical purpose. The post office doesn't accept the old stamps, the banks don't take old coins, and the vintage cars are no longer allowed on the road. These are all side issues; the attraction is that they are in short supply.

In one study, students were asked to arrange ten posters in order of attractiveness—with the agreement that afterward they could keep one poster as a reward for their participation. Five minutes later, they were told that the poster with the third-highest rating was no longer available. Then they were asked to judge all ten from scratch. The poster that was no longer available was suddenly classified as the most beautiful. In psychology, this phenomenon is called "reactance": When we are deprived of an option, we suddenly deem it more attractive. It is a kind of act of defiance. It is also known as the "Romeo and Juliet effect": Because the love between the tragic Shakespearean teenagers is forbidden, it knows no bounds. This yearning must not necessarily be in a romantic way. In the United States, student parties are often littered with desperately drunk teenagers. In Europe, where the age limit is eighteen, you don't witness this type of behavior.

In conclusion: The typical response to scarcity is a lapse in clear thinking. Assess products and services solely on the basis of their price and benefits. It should be of no importance if an item is disappearing fast or if any doctors from London take an interest.

28

When You Hear Hoofbeats, Don't Expect a Zebra
Base-Rate Neglect

Mark is a thin man from Germany with glasses who likes to listen to Mozart. Which is more likely? That (a) Mark is a truck driver or (b) he is a professor of literature in Frankfurt. Most will bet on B, which is wrong. Germany has ten thousand times more truck drivers than Frankfurt has literature professors. Therefore, it is more likely that Mark is a truck driver. So what just happened? The detailed description enticed us to overlook the statistical reality. Scientists call this fallacy *base-rate neglect*: a disregard of fundamental distribution levels. It is one of the most common errors in reasoning. Virtually all journalists, economists, and politicians fall for it on a regular basis.

Here is a second example: A young man is stabbed and fatally injured. Which of these is more likely? (a) The attacker is a Russian immigrant and imports combat knives illegally, or (b) the attacker is a middle-class American. You know the drill now: Option B is much more likely because there are a million times more middle-class Americans than there are Russian knife importers.

In medicine, *base-rate neglect* plays an important role. For example, migraines can point (among others) to a viral infection or a brain tumor. However, viral infections are much more common (in other words, they have a higher base rate), so doctors assess patients for these first before testing for tumors. This is very reasonable. In medical school, residents spend a lot of time purging *base-rate neglect*. The motto drummed into any prospective doctor in the United States is: "When you hear hoofbeats behind you, don't expect to see a zebra," which means: Investigate the most likely ailments before you start diagnosing exotic diseases, even if you are a specialist in that. Doctors are the only professionals who enjoy this base-rate training.

Regrettably, few people in business are exposed to base-rate training. Now and then I see high-flying entrepreneurs' business plans and get very excited by their products, ideas, and personalities. I often catch myself thinking: This could be the next Google! But a glance at the base rate brings me back down to earth. The probability that a firm will survive the first five years is 20 percent. So what, then, is the probability that they will grow into a global corporation? Almost zero. Warren Buffett once explained why he does not invest in biotech companies: "How many of these companies make revenues of several hundred million dollars? It simply does not happen. . . . The most likely scenario is that these firms will just hover somewhere in the middle." This is clear base-rate thinking. For most people, *survivorship bias* (chapter 1) is one of the causes for their base-rate neglect. They tend to see only the successful individuals and companies because the unsuccessful cases are not reported (or underreported). This makes them neglect the large part of the "invisible" cases.

Imagine you are sampling wine in a restaurant and have to guess from which country it is. The label of the bottle is covered. If, like me, you are not a wine connoisseur, the only lifeline you have is the base rate. You know from experience that about three-quarters of the wines on the menu are of French origin, so reasonably, you guess France, even if you suspect a Chilean or Californian twist.

Sometimes I have the dubious honor of speaking in front of students of elite business schools. When I ask them about their career prospects, most answer that, in the medium term, they see themselves on the boards of global companies. Years ago, both my fellow students and I gave the same answer. The way I see it, my role is to give students a base-rate crash course: "With a degree from this school, your chance of landing a spot on the board of a Fortune 500 company is less than 0.1 percent. No matter how smart and ambitious you are, the most likely scenario is that you will end up in middle management." With this, I earn shocked looks and tell myself that I have made a small contribution toward mitigating their future midlife crises.

29

Why the "Balancing Force of the Universe" Is Baloney

Gambler's Fallacy

I n the summer of 1913, something incredible happened in Monte Carlo. Crowds gathered around a roulette table and could not believe their eyes. The ball had landed on black twenty times in a row. Many players took advantage of the opportunity and immediately put their money on red. But the ball continued to come to rest on black. Even more people flocked to the table to bet on red. It had to change eventually! But it was black yet again—and again and again. It was not until the twenty-seventh spin that the ball eventually landed on red. By that time, the players had bet millions on the table. In a few spins of the wheel, they were bankrupt.

The average IQ of pupils in a big city is 100. To investigate this, you take a random sample of fifty students. The first child tested has an IQ of 150. What will the average IQ of your fifty students be? Most people guess 100. Somehow, they think that the super-smart student will be balanced out—perhaps by a dismal student with an IQ of 50 or by two below-average students with IQs of 75. But with such a small sample, that is very unlikely. We must expect that the remaining forty-nine students will represent the average of the population, so they will each have an average IQ of 100. Forty-nine times 100 plus one IQ of 150 gives us an average of 101 in the sample.

The Monte Carlo example and the IQ experiment show that people believe in the "balancing force of the universe." This is the *gambler's fallacy*. However, with independent events, there is no harmonizing force at work: A ball cannot remember how many times it has landed on black. Despite this, one of my friends enters the weekly Mega Millions numbers into a spreadsheet, and then plays those that have appeared the least. All this work is for naught. He is another victim of the *gambler's fallacy*.

The following joke illustrates this phenomenon: A mathematician is afraid of flying due to the small risk of a terrorist attack. So, on every flight he takes a bomb with him in his hand luggage. "The probability of having a bomb on the plane is very low," he reasons, "and the probability of having two bombs on the same plane is virtually zero!"

A coin is flipped three times and lands on heads on each occasion. Suppose someone forces you to spend thousands of dollars of your own money betting on the next toss. Would you bet on heads or tails? If you think like most people, you will choose tails, although heads is just as likely. The *gambler's fallacy* leads us to believe that something must change.

A coin is tossed fifty times, and each time it lands on heads. Again, with someone forcing you to bet, do you pick heads or tails? Now that you've seen an example or two, you're wise to the game: You know that it could go either way. But we've just come across another pitfall: the classic *déformation profession-*

nelle (professional oversight; see chapter 92) of mathematicians: Common sense would tell you that heads is the wiser choice, since the coin is obviously loaded.

In chapter 19, we looked at *regression to mean*. An example: If you are experiencing record cold where you live, it is likely that the temperature will return to normal values over the next few days. If the weather functioned like a casino, there would be a 50 percent chance that the temperature would rise and a 50 percent chance that it would drop. But the weather is not like a casino. Complex feedback mechanisms in the atmosphere ensure that extremes balance themselves out. In other cases, however, extremes intensify. For example, the rich tend to get richer. A stock that shoots up creates its own demand to a certain extent, simply because it stands out so much—a sort of reverse compensation effect.

So, take a closer look at the independent and interdependent events around you. Purely independent events really only exist at the casino, in the lottery, and in theory. In real life, in the financial markets and in business, with the weather and your health, events are often interrelated. What has already happened has an influence on what will happen. As comforting an idea as it is, there is simply no balancing force out there for independent events. "What goes around, comes around" simply does not exist.

30

Why the Wheel of Fortune Makes Our Heads Spin

The Anchor

When was Abraham Lincoln born? If you don't know the year off the top of your head, and your smartphone battery has just died, how do you answer this? Perhaps you know that he was president during the Civil War in the 1860s and that he was the first U.S. president to be assassinated. Looking at the Lincoln Memorial in Washington, you don't see a young, energetic man but something more akin to a worn-out sixty-year-old veteran. The memorial must depict him at the height of his political power, say, at the age of sixty. Let's assume that he was assassinated in the mid-1860s, making 1805 our estimate for the year he was born. (The correct answer is 1809.) So how did we work it out? We found an *anchor* to help us—the year 1865—and worked from there to an educated guess.

Whenever we have to guess something—the length of the Mississippi River, population density in Russia, the number of nuclear power plants in France—we use *anchors*. We start with something we are sure of and venture into unfamiliar territory from there. How else could we do it? Just pick a number off the top of our heads? That would be irrational.

Unfortunately, we also use *anchors* when we don't need to. For example, one day in a lecture, a professor placed a bottle of wine on the table. He asked his students to write down the last two digits of their Social Security numbers and then decide if they would be willing to spend that amount on the wine. In the auction that followed, students with higher numbers bid nearly twice as much as students with lower numbers. The Social Security digits worked as an *anchor*—albeit in a hidden and misleading way.

The psychologist Amos Tversky conducted an experiment involving a wheel of fortune. He had participants spin it, and afterward they were asked how many member states the United Nations has. Their guesses confirmed the *anchor* effect: The highest estimates came from people who had spun high numbers on the wheel.

Researchers Russo and Shoemaker asked students in what year Attila the Hun suffered his crushing defeat in Europe. Just like the example with Social Security numbers, the participants were anchored—this time with the last few digits of their telephone number. The result? People with higher numbers chose later years and vice versa. (If you were wondering, Attila's demise came about in 453.)

Another experiment: Students and professional real estate agents were given a tour of a house and asked to estimate its value. Beforehand, they were informed about a (randomly generated) listed sales price. As might be expected, the *anchor* influenced the students: The higher this price, the higher they valued the property. And the professionals? Did they value the house objectively? No, they were similarly influenced by the random *anchor* amount. The more uncertain the value of something—such as real estate, company stock, or art—the more susceptible even experts are to *anchors*.

Anchors abound, and we all clutch at them. The "recommended retail price" printed on many products is nothing more than an *anchor*. Sales professionals know they must establish a price at an early stage—long before they have an offer. Also, it has been proven that if teachers know students' past grades, it influences how they will mark new work. The most recent grades act as a starting point.

In my early years, I had a quick stint at a consulting firm. My boss was a pro when it came to using *anchors*. In his first conversation with any client, he made sure to fix an opening price, which, by the way, almost criminally exceeded our internal costs: "I'll tell you this now so you're not surprised when you receive the quote, Mr. So-and-So: We've just completed a similar project for one of your competitors and it was in the range of five million dollars." The *anchor* was dropped: The price negotiations started at exactly five million.

31

How to Relieve People of Their Millions
Induction

A farmer feeds a goose. At first, the shy animal is hesitant, wondering: "What's going on here? Why is he feeding me?" This continues for a few more weeks until, eventually, the goose's skepticism gives way. After a few months, the goose is sure: "The farmer has my best interests at heart." Each additional day's feeding confirms this. Fully convinced of the man's benevolence, the goose is amazed when he takes it out of its enclosure on Christmas Day—and slaughters it. The Christmas goose fell victim to *inductive thinking*, the inclination to draw universal certainties from individual observations. Philosopher David Hume used this allegory back in the eighteenth century to warn of its pitfalls. However, it's not just geese that are susceptible to it.

An investor buys shares in stock X. The share price rockets, and at first he is wary. "Probably a bubble," he suspects. As the stock continues to rise, even after months, his apprehension turns into excitement: "This stock may never come down," especially since every day this is the case. After half a year, he invests his life savings in it, turning a blind eye to the huge cluster risk this poses. Later, the man will pay for his foolish investment. He has fallen hook, line, and sinker for *induction*.

Inductive thinking doesn't have to be a road to ruin, though. In fact, you can make a fortune with it by sending a few e-mails. Here's how: Put together two stock market forecasts—one predicting that prices will rise next month and one warning of a drop. Send the first mail to fifty thousand people and the second mail to a different set of fifty thousand. Suppose that after one month, the indices have fallen. Now you can send another e-mail, but this time only to the fifty thousand people who received a correct prediction. These fifty thousand you divide into two groups: The first half learns that prices will increase next month, and the second half discovers they will fall. Continue doing this. After ten months, around a hundred people will remain, all of whom you have advised impeccably. From their perspective, you are a genius. You have proven that you are truly in possession of prophetic powers. Some of these people will trust you with their money. Take it and start a new life in Brazil.

However, it's not just naive strangers who get deceived in this way; we constantly trick ourselves, too. For example, people who are rarely ill consider themselves immortal. CEOs who announce increased profits in consecutive quarters deem themselves infallible—their employees and shareholders do, too. I once had a friend who was a base jumper. He jumped off cliffs, antennae, and buildings, pulling the rip cord only at the last minute. One day, I brought up how risky his chosen sport is. He replied quite matter-of-factly: "I've over a thousand jumps under my belt, and nothing has ever happened to me." Two months later, he was dead. It happened when he jumped from a particularly dangerous cliff in South Africa. This single event was enough to eradicate a theory confirmed a thousand times over.

Inductive thinking can have devastating results. Yet we cannot do without it. We trust that, when we board a plane, aerodynamic laws will still be valid. We imagine that we will not be randomly beaten up on the street. We expect that our hearts will still be beating tomorrow. These are confidences without which we could not live, but we must remember that certainties are always provisional. As Benjamin Franklin said, "Nothing is certain but death and taxes."

Induction seduces us and leads us to conclusions such as: "Mankind has always survived, so we will be able to tackle any future challenges, too." Sounds good in theory, but what we fail to realize is that such a statement can only come from a species that has lasted until now. To assume that our existence to date is an indication of our future survival is a serious flaw in reasoning. Probably the most serious of all.

32

Why Evil Is More Striking Than Good

Loss Aversion

On a scale of 1 to 10, how good do you feel today? Now consider what would bring you up to a perfect 10. That vacation in the Caribbean you've always dreamed of? A step up the career ladder, maybe? Next question: What would make you drop down by the same number of points? Paralysis, Alzheimer's, cancer, depression, war, hunger, torture, financial ruin, damage to your reputation, losing your best friend, your children getting kidnapped, blindness, death? The long list of possibilities makes us realize just how many obstacles to happiness exist; in short, there are more bad things than good—and they are far more consequential.

In our evolutionary past, this was even more the case. One stupid mistake and you were dead. Everything could lead to your rapid departure from the game of life—carelessness on the hunt, an inflamed tendon, exclusion from the group, and so on. People who were reckless or gung ho died before they could pass their genes on to the next generation. Those who remained, the cautious, survived. We are their descendants.

So, no wonder we fear loss more than we value gain. Losing $100 costs you a greater amount of happiness than the delight you would feel if I gave you $100. In fact, it has been proven that, emotionally, a loss "weighs" about twice that of a similar gain. Social scientists call this *loss aversion*.

For this reason, if you want to convince someone about something, don't focus on the advantages; instead highlight how it helps them dodge the disadvantages. Here is an example from a campaign promoting breast self-examination (BSE): Two different leaflets were handed out to women. Pamphlet A urged: "Research shows that women who *do* BSE have an *increased* chance of finding a tumor in the early, more treatable state of the disease." Pamphlet B said: Research shows that women who do *not do* BSE have a *decreased* chance of finding a tumor in the early, more treatable state of the disease." The study revealed that pamphlet B (written in a "loss frame") generated significantly more awareness and BSE behavior than pamphlet A (written in a "gain frame").

The fear of losing something motivates people more than the prospect of gaining something of equal value. Suppose your business is home insulation. The most effective way of encouraging customers to purchase your product is to tell them how much money they are losing without insulation—as opposed to how much money they would save with it, even though the amount is exactly the same.

This type of aversion is also found on the stock market, where investors tend to simply ignore losses on paper. After all, an unrealized loss isn't as painful as a realized one. So they sit on the stock, even if the chance of recovery is small and the probability of further decline is large. I once met a man, a multimillionaire, who was terribly upset because he had lost a $100 bill. What a waste of emotion! I pointed out that the value of his portfolio fluctuated by at least $100 every second.

Management gurus push employees in large companies to be bolder and more entrepreneurial. The reality is: Employees tend to be risk averse. From their perspective, this aversion makes perfect sense: Why risk something that brings them, at best, a nice bonus, and at worst, a pink slip? The downside is larger than the upside. In almost all companies and situations, safeguarding your career trumps any potential reward. So, if you've been scratching your head about the lack of risk taking among your employees, you now know why. (However, if employees do take big risks, it is often when they can hide behind group decisions. Learn more in chapter 33 on *social loafing*.)

We can't fight it: Evil is more powerful and more plentiful than good. We are more sensitive to negative than to positive things. On the street, scary faces stand out more than smiling ones. We remember bad behavior longer than good—except, of course, when it comes to ourselves.

33

Why Teams Are Lazy
Social Loafing

Maximilian Ringelmann, a French engineer, studied the performance of horses in 1913. He concluded that the power of two animals pulling a coach did not equal twice the power of a single horse. Surprised by this result, he extended his research to humans. He had several men pull a rope and measured the force applied by each individual. On average, if two people were pulling together, each invested just 93 percent of his individual strength, when three pulled together, it was 85 percent, and with eight people, just 49 percent.

Science calls this the *social loafing* effect. It occurs when individual performance is not directly visible; it blends into the group effort. It occurs among rowers, but not in relay races, because here, individual contributions are evident. *Social loafing* is rational behavior: Why invest all of your energy when half will do—especially when this little shortcut goes unnoticed? Quite simply, *social loafing* is a form of cheating of which we are all guilty even if it takes place unconsciously, just as it does with the horses.

When people work together, individual performances decrease. This isn't surprising. What is noteworthy, however, is that our input doesn't grind to a complete halt. So what stops us from putting our feet up and letting the others do the hard work? The consequences. Zero performance would be noticed, and it brings with it weighty punishments, such as exclusion from the group or vilification. Evolution has led us to develop many fine-tuned senses, including how much idleness we can get away with and how to recognize it in others.

Social loafing does not occur solely in physical performance. We slack off mentally, too. For example, in meetings, the larger the team, the weaker our individual participation. However, once a certain number of participants are involved, our performance plateaus. Whether the group consists of twenty or one hundred people is not important—maximum inertia has been achieved.

One question remains: Who came up with the much-vaunted idea that teams achieve more than individual workers? Maybe the Japanese. Thirty years ago, they flooded global markets with their products. Business economists looked more closely at the industrial miracle and saw that Japanese factories were organized into teams. This model was copied—with mixed success. What worked very well in Japan could not be replicated with the Americans and Europeans—perhaps because *social loafing* rarely happens there. In the West, teams function better if and only if they are small and consist of diverse, specialized people. This makes sense, because within such groups, individual performances can be traced back to each specialist.

Social loafing has interesting implications. In groups, we tend to hold back not only in terms of participation but also in terms of accountability. Nobody wants to take the rap for the misdeeds or poor decisions of the whole group. A glaring example is the prosecution of the Nazis at the Nuremberg trials or, less controversially, any board or management team. We hide behind team decisions. The technical term for this is "diffusion of responsibility." For the same reason, teams tend to take bigger risks than their members would take on their own. The individual group members reason that they are not the only ones who will be blamed if things go wrong. This effect is called "risky shift" and is especially hazardous among company and pension-fund strategists, where billions are at stake, or in the Defense Department, where groups decide on the use of nuclear weapons.

In conclusion: People behave differently in groups than when alone (otherwise there would be no groups). The disadvantages of groups can be mitigated by making individual performances as visible as possible. Long live meritocracy! Long live the performance society!

34

Stumped by a Sheet of Paper

Exponential Growth

A piece of paper is folded in two, then in half again, and again and again. How thick will it be after fifty folds? Write down your guess before you continue reading.

Second task. Choose between these options: (a) Over the next thirty days, I will give you $1,000 a day. (b) Over the next thirty days, I will give you a cent on the first day, two cents on the second day, four cents on the third day, eight cents on the fourth day, and so on. Don't think too long about it: A or B?

Are you ready? Well, if we assume that a sheet of copy paper is approximately 0.004 inches thick, then its thickness after fifty folds is a little over sixty million miles. This equals the distance between the earth and the sun, as you can check easily with a calculator. With the second question, it is worthwhile choosing option B, even though A sounds more tempting. Selecting A earns you $30,000 in thirty days; choosing B gives you more than $5 million.

Linear growth we understand intuitively. However, we have no sense of exponential (or percentage) growth. Why is this? Because we didn't need it before. Our ancestors' experiences were mostly of the linear variety. Whoever spent twice the time collecting berries earned double the amount. Whoever hunted two mammoths instead of one could eat for twice as long. In the Stone Age, people rarely came across exponential growth. Today, things are different.

"Each year, the number of traffic accidents rises by 7 percent," warns a politician. Let's be honest: We don't intuitively understand what this means. So, let's use a trick and calculate the "doubling time." Start with the magic number of 70 and divide it by the growth rate in percent. In this instance: 70 divided by 7 = 10 years. So what the politician is saying is: "The number of traffic accidents doubles every ten years." Pretty alarming. (You may ask: "Why the number 70?" This has to do with a mathematical concept called logarithm. You can look it up in the notes section.)

Another example: "Inflation is at 5 percent." Whoever hears this thinks: "That's not so bad, what's 5 percent anyway?" Let's quickly calculate the doubling time: 70 divided by 5 = 14 years. In fourteen years, a dollar will be worth only half what it is today—a catastrophe for anyone who has a savings account.

Suppose you are a journalist and learn that the number of registered dogs in your city is rising by 10 percent a year. Which headline do you put on your article? Certainly not: "Dog Registrations Increasing by 10 Percent." No one will care. Instead, announce: "Deluge of Dogs: Twice as Many Mutts in Seven Years' Time!"

Nothing that grows exponentially grows forever. Most politicians, economists, and journalists forget that. Such growth will eventually reach a limit. Guaranteed. For example, the intestinal bacterium *Escherichia coli* divides every twenty minutes.

In just a few days, it could cover the whole planet, but since it consumes more oxygen and sugar than is available, its growth has a cutoff point.

The ancient Persians were well aware that people struggled with percentage growth. Here is a local tale: There was once a wise courtier who presented the king with a chessboard. Moved by the gift, the king said to him: "Tell me how I can thank you." "Your highness, I want nothing more than for you to cover the chess board with rice, putting one grain of rice on the first square, and then on every subsequent square, twice the previous number of grains." The king was astonished: "It is an honor to you, dear courtier, that you present such a modest request." But how much rice is that? The king guessed about a sack. Only when his servants began the task—placing a grain on the first square, two grains of rice on the second square, four grains of rice on the third, and so on—did he realize that he would need more rice than was growing on earth.

When it comes to growth rates, do not trust your intuition. You don't have any. Accept it. What really helps is a calculator or, with low growth rates, the magic number of 70.

35

Curb Your Enthusiasm

Winner's Curse

T exas in the 1950s. A piece of land is being auctioned. Ten oil companies are vying for it. Each has made an estimate of how much the site is worth. The lowest assessment is $10 million, and the highest is $100 million. The higher the price climbs during the auction, the more firms exit the bidding. Finally, one company submits the highest bid and wins. Champagne corks pop.

The *winner's curse* suggests that the winner of an auction often turns out to be the loser. Industry analysts have noted that companies that regularly emerged as winning bidders from these oil field auctions systematically paid too much and years later went under. This is understandable. If the estimates vary between $10 million and $100 million, the actual value most likely lies somewhere in the middle. The highest bid at an auction is often much too high—unless these bidders have critical information others are not privy to. This was not the case in Texas. The oil managers actually celebrated a Pyrrhic victory.

Today, this phenomenon affects us all. From eBay to Groupon to Google AdWords, prices are consistently set by auction.

Bidding wars for cell-phone frequencies drive telecom companies to the brink of bankruptcy. Airports rent out their commercial spaces to the highest bidder. And if Walmart plans to introduce a new detergent and asks for tenders from five suppliers, that's nothing more than an auction—with the risk of the *winner's curse*.

The auctioning of everyday life has now reached tradesmen, too, thanks to the Internet. When my walls needed a new lick of paint, instead of tracking down the handiest painter, I advertised the job online. Thirty painters from more than three hundred miles away competed for the job. The best offer was so low that, out of compassion, I could not accept it—to spare the poor painter the *winner's curse*.

Initial public offerings (IPOs) are also examples of auctions. And when companies buy other companies—the infamous mergers and acquisitions—the *winner's curse* is present more often than not. Astoundingly, more than half of all acquisitions destroy value, according to a McKinsey study.

So why do we fall victim to the *winner's curse*? First, the real value of many things is uncertain. Additionally, the more interested parties, the greater the likelihood of an overly enthusiastic bid. Second, we want to outdo competitors. A friend owns a micro-antenna factory and told me about the cutthroat bidding war that Apple instigated during the development of the iPhone. Everyone wants to be the official supplier to Apple, even though whoever gets the contract is likely to lose money.

So how much would you pay for $100? Imagine that you and an opponent are invited to take part in such an auction. The rules: Whoever makes the highest offer gets the $100 bill, and—most important—when this happens, both bidders have to pay their final offer. How high will you go? From your perspective, it makes sense to pay $20, $30, or $40. Your opponent does the same. Even $99 seems like a reasonable offer for a $100 bill. Now, your competitor offers $100. If this remains the highest bid, he will come away breaking even (paying $100 for $100), whereas you will simply have to cough up $99. So you continue to bid. At $110, you have a guaranteed loss of $10, but your opponent would have to shell out $109 (his last bid). So he will continue playing. When do you stop? When will your competitor give up? Try it out with friends.

In conclusion: Accept this piece of wisdom about auctions from Warren Buffett: "Don't go." If you happen to work in an industry where they are inevitable, set a maximum price and deduct 20 percent from this to offset the *winner's curse*. Write this number on a piece of paper and don't go a cent over it.

36

Never Ask a Writer If the Novel Is Autobiographical

Fundamental Attribution Error

Opening the newspaper, you learn that another CEO has been forced to step down because of bad results. In the sports section, you read that your team's winning season was thanks to player X or coach Y. In history books, you learn that the success of the French army in the early 1800s is a testament to Napoleon's superb leadership and strategy. "Every story has a face," it seems. Indeed, this is an ironclad rule in every newsroom. Always on the lookout for the "people angle," journalists (and their readers) take this principle one step further, and thus fall prey to the *fundamental attribution error*. This describes the tendency to overestimate individuals' influence and underestimate external, situational factors.

In 1967, researchers at Duke University set up the following experiment: Participants read an argument either lauding or loathing Fidel Castro. They were informed that the author of the text had been allocated the viewpoint regardless of his true political views; he was just making a coherent argument. Nevertheless, most of the audience believed what he said reflected his true opinion. They falsely attributed the content of the speech to his character and ignored the external factors—in this case, the professors who had crafted the text.

The *fundamental attribution error* is particularly useful for whittling negative events into neat little packages. For example, the "blame" for wars we lazily push onto individuals: The Yugoslav assassin in Sarajevo has World War I on his conscience, and Hitler singlehandedly caused World War II. Many swallow these simplifications, even though wars are unforeseeable events whose innumerable dynamics we may never fully understand. Which sounds a little like financial markets and climate issues, don't you agree?

We see this same pattern when companies announce good or bad results. All eyes shift to the CEO's office, even if we know the truth: Economic success depends far more on the overall economic climate and the industry's attractiveness than on brilliant leadership. It is interesting how frequently firms in ailing industries replace their CEOs—and how seldom that happens in booming sectors. Are ailing industries less careful in their recruitment processes? Such decisions are no more rational than what happens between football coaches and their clubs.

I often go to musical concerts. In my hometown of Lucerne, in the center of Switzerland, I am spoiled with one-off classical recitals. During the intermission, however, I notice that the conversations almost always revolve around the conductors and/or soloists. With the exception of world premieres, composition is rarely discussed. Why not? The real miracle of music is, after all, the composition, the creation of sounds, moods, and rhythms where previously only a blank sheet lay. The difference among scores is a thousand times more impressive than the difference among performances of the same score. But we do not think like this. The score is—in contrast to the conductors and soloists—faceless.

In my career as a fiction writer, I experience the *fundamental attribution error* in this way: After a reading (which in itself is a debatable undertaking), the first question always, really always, is: "What part of your novel is autobiographical?" I often feel like thundering: "It's not about me, damn it! It's about the book, the text, the language, the credibility of the story!" But unfortunately my upbringing allows such outbursts only rarely.

We shouldn't judge those guilty of the *fundamental attribution error* too harshly. Our preoccupation with other people stems from our evolutionary past: Belonging to a group was necessary for survival. Reproduction, defense, and hunting large animals—all these were impossible tasks for individuals to achieve alone. Banishment meant certain death, and those who opted for the solitary life—of which there were surely a few—fared no better and disappeared from the gene pool. In short, our lives depended on and revolved around others, which explains why we are so obsessed with our fellow humans today. The result of this infatuation is that we spend about 90 percent of our time thinking about other people and dedicate just 10 percent to assessing other factors and contexts.

In conclusion: As much as we are fascinated with the spectacle of life, the people onstage are not perfect, self-governed individuals. Instead, they tumble from situation to situation. If you want to understand the current play—really understand it—then forget about the performers. Pay close attention to the dance of influences to which the actors are subjected.

37

Why You Shouldn't Believe in the Stork
False Causality

For the inhabitants of the Hebrides, a chain of islands north of Scotland, head lice were a part of life. If the lice left their host, he became sick and feverish. Therefore, to dispel the fever, sick people had lice put in their hair intentionally. There was a method to their madness: As soon as the lice had settled in again, the patient improved.

In one city, a study revealed that in each blaze, the more firefighters called out to fight it, the greater the fire damage. The mayor imposed an immediate hiring freeze and cut the firefighting budget.

Both stories come from German physics professors Hans-Peter Beck-Bornholdt and Hans-Hermann Dubben. In their book (unfortunately there is no English version), they illustrate the muddling of cause and effect. If the lice leave the invalid, it is because he has a fever and they simply get hot feet. When the fever breaks, they return. And the bigger the blaze, the more firefighters were called out—not, of course, vice versa.

We may smirk at these stories, but *false causality* leads us astray practically every day. Consider the headline: "Employee Motivation Leads to Higher Corporate Profits." Does it? Maybe people are simply more motivated because the company is doing well. Another headline touts that the more women on a corporate board, the more profitable the firm is. But is that really how it works? Or do highly profitable firms simply tend to recruit more women to their boards? Business book authors and consultants often operate with similar false—or at least fuzzy—causalities.

In the '90s, there was no one holier than the then-head of the Federal Reserve, Alan Greenspan. His obscure remarks gave monetary policy the aura of a secret science that kept the country on the secure path of prosperity. Politicians, journalists, and business leaders idolized Greenspan. Today we know that these commentators fell victim to *false causality*. America's symbiosis with China, the globe's low-cost producer and eager buyer of U.S. debt, played a much more important role. In other words, Greenspan was simply lucky that the economy did so well during his tenure.

A further example: Scientists found that long periods in the hospital affected patients adversely. This was music to health insurers' ears, who, of course, are keen to make stays as brief as possible. But, clearly, patients who are discharged immediately are healthier than those who must stay on for treatment. This hardly makes long stays detrimental.

Or, take this headline: "Fact: Women Who Use Shampoo XYZ Every Day Have Stronger Hair." Though the context can be substantiated scientifically, this statement says very little— least of all, that the shampoo makes your hair stronger. It might simply be the other way round: Women with strong hair tend to use shampoo XYZ—and perhaps that's because it says "especially for thick hair" on the bottle.

Recently I read that students get better grades at school if their homes contain a lot of books. This study was surely a shot in the arm for booksellers, but it is another fine example of *false causality*. The simple truth is that educated parents tend to value their children's education more than uneducated ones do. Plus, educated parents often have more books at home. In short, a dust-covered copy of *War and Peace* alone isn't going to influence anyone's grades; what counts is parents' education levels, as well as their genes.

The best example of *false causality* was the supposed relationship between the birth rate and the numbers of stork pairs in Germany. Both were in decline, and if you plot them on a graph, the two lines of development from 1965 to 1987 appeared almost identical. Does this mean the stork actually does bring babies? Obviously not, since this was a purely coincidental correlation.

In conclusion: Correlation is not causality. Take a closer look at linked events: Sometimes what is presented as the cause turns out to be the effect, and vice versa. And sometimes there is no link at all—just like with the storks and babies.

38

Why Attractive People Climb the Career Ladder More Quickly

Halo Effect

Cisco, the Silicon Valley firm, was once a darling of the new economy. Business journalists gushed about its success in every discipline: its wonderful customer service, perfect strategy, skillful acquisitions, unique corporate culture, and charismatic CEO. In March 2000, it was the most valuable company in the world.

When Cisco's stock plummeted 80 percent the following year, the journalists changed their tune. Suddenly the company's competitive advantages were reframed as destructive shortcomings: poor customer service, a woolly strategy, clumsy acquisitions, a lame corporate culture, and an insipid CEO. All this—and yet neither the strategy nor the CEO had changed. What had changed, in the wake of the dot-com crash, was demand for Cisco's product—and that was through no fault of the firm.

The *halo effect* occurs when a single aspect dazzles us and affects how we see the full picture. In the case of Cisco, its halo shone particularly bright. Journalists were astounded by its stock prices and assumed the entire business was just as brilliant—without closer investigation.

The *halo effect* always works the same way: We take a simple-to-obtain or remarkable fact or detail, such as a company's financial situation, and extrapolate conclusions from there that are harder to nail down, such as the merit of its management or the feasibility of its strategy. We often ascribe success and superiority where little is due, such as when we favor products from a manufacturer simply because of its good reputation. Another example of the *halo effect*: We believe that CEOs who are successful in one industry will thrive in any sector—and furthermore that they are heroes in their private lives, too.

The psychologist Edward Lee Thorndike discovered the *halo effect* nearly one hundred years ago. His conclusion was that a single quality (e.g., beauty, social status, age) produces a positive or negative impression that outshines everything else, and the overall effect is disproportionate. Beauty is the best-studied example. Dozens of studies have shown that we automatically regard good-looking people as more pleasant, honest, and intelligent. Attractive people also have it easier in their professional lives—and that has nothing to do with the myth of (women) "sleeping their way to the top." The effect can even be detected in schools, where teachers unconsciously give good-looking students better grades.

Advertising has found an ally in the *halo effect*: Just look at the number of celebrities smiling at us from TV ads, billboards, and magazines. What makes a professional tennis player like Roger Federer a coffee machine expert is still open for debate, but this hasn't detracted from the success of the campaign. We are so used to seeing celebrities promoting arbitrary products that we never stop to consider why their support should be of any importance to us. But this is exactly the sneaky part of the *halo effect*: It works on a subconscious level. All that needs to register is the attractive face, dream lifestyle—and that product.

Sticking with negative effects, the *halo effect* can lead to great injustice and even stereotyping when nationality, gender, or race becomes the all-encompassing feature. One need be neither racist nor sexist to fall victim to this. The *halo effect* clouds our view, just as it does journalists, educators, and consumers.

Occasionally, this effect has pleasant consequences—at least in the short term. Have you ever been head over heels in love? If so, you know how flawless a person can appear. Your Mr. or Ms. Perfect seems to be the whole package: attractive, intelligent, likable, and warm. Even when your friends might point out obvious failings, you see nothing but endearing quirks.

The *halo effect* obstructs our view of true characteristics. To counteract this, go beyond face value. Factor out the most striking features. World-class orchestras achieve this by making candidates play behind a screen, so that sex, race, age, and appearance play no part in their decision. To business journalists I warmly recommend judging a company by something other than its easily obtainable quarterly figures (the stock market already delivers that). Dig deeper. Invest the time to do serious research. What emerges is not always pretty, but almost always educational.

Congratulations! You've Won Russian Roulette
Alternative Paths

You arrange to meet with a Russian oligarch in a forest just outside your city. He arrives shortly after you, carrying a suitcase and a gun. Placing the suitcase on the hood of his car, he opens it so you can see it is filled to the brim with stacks of money—$10 million in total. "Want to play Russian roulette?" he asks. "Pull the trigger once, and all this is yours." The revolver contains a single bullet; the other five chambers are empty. You consider your options. Ten million dollars would change your life. You would never have to work again. You could finally move from collecting stamps to collecting sports cars!

You accept the challenge. You put the revolver to your temple and squeeze the trigger. You hear a faint click and feel adrenaline flood your body. Nothing happens. The chamber was empty! You have survived. You take the money, move to the most beautiful city you know, and upset the locals by building a luxurious villa there.

One of these neighbors, whose home now stands in the shadow of yours, is a prominent lawyer. He works twelve hours a day, three hundred days a year. His rates are impressive, but not unusual: $500 per hour. Each year he can put aside half a million dollars net after taxes and living expenses. From time to time, you wave to him from your driveway, laughing on the inside: He will have to work for twenty years to catch up with you.

Suppose that, after twenty years, your hardworking neighbor has saved up $10 million. A journalist comes along one day and puts together a piece on the more affluent residents in the area—complete with photos of the magnificent buildings and the beautiful second wives that you and your neighbor have accrued. He comments on the interior design and the exquisite landscaping. However, the crucial difference between the two of you remains hidden from view: the risk that lurks behind each of the $10 million. For this he would need to recognize the *alternative paths*.

But not only journalists are underachievers at this skill. We all are.

Alternative paths are all the outcomes that could have happened but did not. With the game of Russian roulette, four *alternative paths* would have led to the same result (winning the $10 million) and the fifth alternative to your death. A huge difference. In the case of the lawyer, the possible paths lie much more closely together. In a village, he would have earned perhaps just $200 per hour. In the heart of New York working for one of the major investment banks, maybe it would have been $600 per hour. But, unlike you, he risked no *alternative path* that would have cost him his fortune—or his life.

Alternative paths are invisible, so we contemplate them very rarely. Those who speculate on junk bonds, options, and credit default swaps, thus making millions, should never forget that they flirt with many *alternative paths* that lead straight to ruin. To a rational mind, $10 million that comes about through a huge risk is worth less than the same sum earned by years of drudgery. (An accountant might disagree, though.)

Recently, I was at a dinner with an American friend who suggested tossing a coin to decide who should pay the bill. He lost. The situation was uncomfortable for me, since he was my guest in Switzerland. "Next time I'll pay, whether here or in New York," I promised. He thought for a moment and said, "Considering the *alternative paths*, you've actually already paid for half of this dinner."

In conclusion: Risk is not directly visible. Therefore, always consider what the *alternatives paths* are. Success that comes about through risky dealings is, to a rational mind, of less worth than success achieved the "boring" way (for example, with laborious work as a lawyer, a dentist, a ski instructor, a pilot, a hairdresser, or a consultant). Yes, looking at *alternative paths* from the outside is a difficult task, looking at them from the inside an almost impossible task. Your brain will do everything to convince you that your success is warranted—no matter how risky your dealings are—and will obscure any thought of paths other than the one you are on.

40

False Prophets

Forecast Illusion

F acebook to be number one entertainment platform in three years."

"Regime shift in North Korea in two years."

"Sour grapes for France as Argentinian wines expected to dominate."

"Euro collapse likely."

"Low-cost space flights by 2025."

"No more crude oil in fifteen years."

Every day, experts bombard us with predictions, but how reliable are they? Until a few years ago, no one bothered to check. Then along came Philip Tetlock. Over a period of ten years, he evaluated 28,361 predictions from 284 self-appointed professionals. The result: In terms of accuracy, the experts fared only marginally better than a random forecast generator. Ironically, the media darlings were among the poorest performers; and of those, the worst were the prophets of doom and disintegration. Examples of their far-fetched forecasts included the collapse of Canada, Nigeria, China, India, Indonesia, South Africa, Belgium, and the EU. None of these countries has imploded.

"There are two kinds of forecasters: those who don't know, and those who don't know they don't know," wrote Harvard economist John Kenneth Galbraith. With this he made himself a figure of hatred in his own guild. Fund manager Peter Lynch summed it up even more cuttingly: "There are 60,000 economists in the U.S., many of them employed full-time trying to forecast recessions and interest rates, and if they could do it successfully twice in a row, they'd all be millionaires by now. . . . As far as I know, most of them are still gainfully employed, which ought to tell us something." That was ten years ago. Today, the United States could employ three times as many economists—with little or no effect on the quality of their forecasts.

The problem is that experts enjoy free rein with few negative consequences. If they strike it lucky, they enjoy publicity, consultancy offers, and publication deals. If they are completely off the mark, they face no penalties—neither in terms of financial compensation nor in loss of reputation. This win-win scenario virtually incentivizes them to churn out as many prophecies as they can muster. Indeed, the more forecasts they generate, the more will be coincidentally correct. Ideally, they should have to pay into some sort of "forecast fund"—say, $1,000 per prediction. If the forecast is correct, the expert gets his money back with interest. If he is wrong, the money goes to charity.

So what is predictable and what is not? Some things are fairly simple. For example, I have a rough idea of how many pounds I will weigh in a year's time. However, the more complex a system, and the longer the time frame, the more blurred the view of the future will be. Global warming, oil prices, or exchange rates are almost impossible to foresee. Inventions are not at all predictable because if we knew what technology we would invent in the future, we would already have invented it.

So, be critical when you encounter predictions. Whenever I hear one, I make sure to smile, no matter how bleak it is. Then I ask myself two questions. First, what incentive does the expert have? If he is an employee, could he lose his job if he is always wrong? Or is he a self-appointed guru who earns a living through books and lectures? The latter type of forecaster relies on the media's attention so, predictably, his prophecies tend to be sensational. Second, how good is his success rate? How many predictions has he made over the past five years? Out of these, how many have been right and how many have not? This information is vital, yet often goes unreported. I implore the media: Please don't publish any more forecasts without giving the pundit's track record.

Finally, since it is so fitting, a quote from former British prime minister Tony Blair: "I don't make predictions. I never have, and I never will."

41

The Deception of Specific Cases

Conjunction Fallacy

Chris is thirty-five. He studied social philosophy and has had an interest in developing countries since he was a teenager. After graduation, he worked for two years with the Red Cross in West Africa and then for three years in its Geneva headquarters, where he rose to head of the African aid department. He then completed an MBA, writing his thesis on corporate social responsibility. What is more likely? (a) Chris works for a major bank or (b) Chris works for a major bank, where he runs its Third World foundation. A or B?

Most people will opt for B. Unfortunately, it's the wrong answer. Option B does not only say that Chris works for a major bank but also that an additional condition has been met. Employees who work specifically within a bank's Third World foundation comprise a tiny subset of bankers. Therefore, option A is much more likely. The *conjunction fallacy* is at play when such a subset seems larger than the entire set—which by definition cannot be the case. Amos Tversky and Nobel laureate Daniel Kahneman have studied this extensively.

We are easy prey for the *conjunction fallacy* because we have an innate attraction to "harmonious" or "plausible" stories. The more convincingly, impressively, or vividly that Chris the aid worker is portrayed, the greater the risk of false reasoning. If I had put it a different way, you would have recognized the extra details as overly specific, for example: "Chris is thirty-five. What is more likely? (a) Chris works for a bank or (b) Chris works for a bank in New York, where his office is on the twenty-fourth floor, overlooking Central Park."

Here's another example: What is more likely? (a) "Seattle airport is closed. Flights are canceled," or (b) "Seattle airport is closed due to bad weather. Flights are canceled." A or B? This time, you have it: A is more likely since B implies that an additional condition has been met, namely, bad weather. It could be that a bomb threat, accident, or strike closed the airport; however, when faced with a "plausible" story, we don't stop to consider such things. Now that you are aware of this, try it out with friends. You will see that most pick B.

Even experts are not immune to the *conjunction fallacy*. In 1982, at an international conference for future research, experts—all of them academics—were divided into two groups. To group A, Daniel Kahneman presented the following forecast for 1983: "Oil consumption will decrease by 30 percent." Group B heard: "A dramatic rise in oil prices will lead to a 30 percent reduction in oil consumption." Both groups had to indicate how likely they considered the scenarios. The result was clear: Group B felt much more strongly about its forecast than group A did.

Kahneman believes that two types of thinking exist: The first kind is intuitive, automatic, and direct. The second is conscious, rational, slow, laborious, and logical. Unfortunately, intuitive thinking draws conclusions long before the conscious mind does. For example, I experienced this after the 9/11 attacks on the World Trade Center. I wanted to take out travel insurance and came across a firm that offered special "terrorism cover." Although other policies protected against all possible incidents (including terrorism), I automatically fell for the offer. The high point of the whole farce was that I was willing to pay even more for this enticing yet redundant add-on.

In conclusion: Forget about left brains and right brains: The difference between intuitive and conscious thinking is much more significant. With important decisions, remember that, at the intuitive level, we have a soft spot for plausible stories. Therefore, be on the lookout for convenient details and happy endings. Remember: If an additional condition has to be met, no matter how plausible it sounds, it will become less, not more, likely.

42

It's Not What You Say, but How You Say It
Framing

C onsider these two statements:
"Hey, the trash can is full!"
"It would be really great if you could empty the trash, honey."

C'est le ton qui fait la musique: it's not what you say but how you say it. If a message is communicated in different ways, it will also be received in different ways. In psychologists' jargon, this technique is called *framing*.

We react differently to identical situations, depending on how they are presented. Kahneman and Tversky conducted a survey in the 1980s in which they put forward two options for an epidemic-control strategy. The lives of six hundred people were at stake, they told participants. "Option A saves two hundred lives. Option B offers a 33 percent chance that all six hundred people will survive, and a 66 percent chance that no one will survive." Although options A and B were comparable (with two hundred survivors expected), the majority of respondents chose A—remembering the adage: A bird in the hand is worth two in the bush. It became really interesting when the same options were *reframed*. "Option A *kills* four hundred people. Option B offers a 33 percent chance that no one will die, and with a 66 percent chance that all six hundred will die." This time, only a fraction of respondents chose A and the majority picked B. The researchers observed a complete U-turn from almost all involved. Depending on the phrasing—survive or die—the respondents made completely different decisions.

Another example: Researchers presented a group of people with two kinds of meat, "99 percent fat free" and "1 percent fat," and asked them to choose which was healthier. Can you guess which they picked? Bingo: Respondents ranked the first type of meat as healthier, even though both were identical. Next came the choice between "98 percent fat free" and "1 percent fat." Again, most respondents chose the first option—despite its higher fat content.

"Glossing" is a popular type of *framing*. Under its rules, a tumbling share price becomes a "correction." An overpaid acquisition price is branded "goodwill." In every management course, a problem magically transforms into an "opportunity" or a "challenge." A person who is fired is "reassessing his career." A fallen soldier—regardless of how much bad luck or stupidity led to his death—turns into a "war hero." Genocide translates to "ethnic cleansing." A successful emergency landing, for example on the Hudson River, is celebrated as a "triumph of aviation." (Shouldn't a textbook landing on a runway count as an even bigger triumph of aviation?)

Have you ever looked more closely at the prospectus for financial products—for example, ETFs (exchange-traded funds)? Generally the brochure illustrates the product's performance in recent years, going back just far enough for the nicest possible upward curve to emerge. This is also *framing*. Another example is a simple piece of bread. Depending on how it is *framed*, as either the "symbolic" or the "true" body of Christ, it can split a religion, as happened in the sixteenth century with the Reformation.

Framing is used to good effect in commerce, too. Consider used cars. You are led to focus on just a few factors, whether the message is delivered through a salesman, a sign touting certain features, or even your own criteria. For example, if the car has the low mileage and good tires, you home in on this and overlook the state of the engine, the brakes, or the interior. Thus, the mileage and tires become the main selling points and frame our decision to buy. Such oversight is only natural, though, since it is difficult to take in all possible pros and cons. Interestingly, had other frames been used to tout the car, we might have decided very differently.

Authors are conscious framers, too. A crime novel would be rather dull if, from page one, the murder were shown as it happened—stab by stab, as it were. Even though we eventually discover the motives and murder weapons, the novelist's *framing* injects thrills and suspense into the story.

In conclusion: Realize that whatever you communicate contains some element of *framing*, and that every fact—even if you hear it from a trusted friend or read it in a reputable newspaper—is subject to this effect, too. Even this chapter.

43
Why Watching and Waiting Is Torture
Action Bias

In a penalty situation in soccer, the ball takes less than 0.3 seconds from the player who kicks the ball to the goal. There is not enough time for the goalkeeper to watch the ball's trajectory. He must make a decision before the ball is kicked. Soccer players who take penalty kicks shoot one third of the time at the middle of the goal, one third of the time at the left, and one third of the time at the right. Surely goalkeepers have spotted this, but what do they do? They dive either to the left or to the right. Rarely do they stay standing in the middle—even though roughly a third of all balls land there. Why on earth would they jeopardize saving these penalties? The simple answer: appearance. It looks more impressive and feels less embarrassing to dive to the wrong side than to freeze on the spot and watch the ball sail past. This is the *action bias*: Look active, even if it achieves nothing.

This study comes from the Israeli researcher Michael Bar-Eli, who evaluated hundreds of penalty shoot-outs. But not just goalkeepers fall victim to the *action bias*. Suppose a group of youths exit a nightclub and begin to argue, shouting at each other and gesturing wildly. The situation is close to escalating into an all-out brawl. The police officers in the area—some young, some more senior—hold back, monitor the scene from a distance, and intervene only when the first casualties appear. If no experienced officers are involved, this situation often ends differently: Young, overzealous officers succumb to the *action bias* and dive in immediately. A study revealed that later intervention, thanks to the calming presence of senior officers, results in fewer casualties.

The *action bias* is accentuated when a situation is new or unclear. When starting out, many investors act like the young, gung ho police officers outside the nightclub: They can't yet judge the stock market so they compensate with a sort of hyperactivity. Of course this is a waste of time. As Charlie Munger sums up his approach to investing: "We've got . . . discipline in avoiding just doing any damn thing just because you can't stand inactivity."

The *action bias* exists even in the most educated circles. If a patient's illness cannot yet be diagnosed with certainty, and doctors must choose between intervening (i.e., prescribing something) or waiting and seeing, they are prone to take action. Such decisions have nothing to do with profiteering, but rather with the human tendency to want to do anything but sit and wait in the face of uncertainty.

So what accounts for this tendency? In our old hunter-gatherer environment (which suited us quite well), action trumped reflection. Lightning-fast reactions were essential to survival; deliberation could be fatal. When our ancestors saw a silhouette appear at the edge of the forest—something that looked a lot like a saber-toothed tiger—they did not take a pew to muse over what it might be. They hit the road—and fast. We are the descendants of these quick responders. Back then, it was better to run away once too often. However, our world today is different; it rewards reflection, even though our instincts may suggest otherwise.

Although we now value contemplation more highly, outright inaction remains a cardinal sin. You get no honor, no medal, no statue with your name on it if you make exactly the right decision by *waiting*—for the good of the company, the state, even humanity. On the other hand, if you demonstrate decisiveness and quick judgment, and the situation improves (though perhaps coincidentally), it's quite possible your boss, or even the mayor, will shake your hand. Society at large still prefers rash action to a sensible wait-and-see strategy.

In conclusion: In new or shaky circumstances, we feel compelled to do something, anything. Afterward we feel better, even if we have made things worse by acting too quickly or too often. So, though it might not merit a parade in your honor, if a situation is unclear, hold back until you can assess your options. "All of humanity's problems stem from man's inability to sit quietly in a room alone," wrote Blaise Pascal. At home, in his study.

44

Why You Are Either the Solution—or the Problem
Omission Bias

You are on a glacier with two climbers. The first slips and falls into a crevasse. He might survive if you call for help, but you don't, and he perishes. The second climber you actively push into the ravine, and he dies shortly afterward. Which weighs more heavily on your conscience?

Considering the options rationally, it's obvious that both are equally reprehensible, resulting as they do in death for your companions. And yet something makes us rate the first option, the passive option, as less horrible. This feeling is called the *omission bias*. It crops up where both action and inaction lead to cruel consequences. In such cases, we tend to prefer inaction; its results seem more anodyne.

Suppose you are the head of the Federal Drug Administration. You must decide whether or not to approve a drug for the terminally ill. The pills can have fatal side effects: They kill 20 percent of patients on the spot, but save the lives of the other 80 percent within a short period of time. What do you decide?

Most would withhold approval. To them, waving through a drug that takes out every fifth person is a worse act than failing to administer the cure to the other 80 percent of patients. It is an absurd decision, and a perfect example of the *omission bias*. Suppose that you are aware of the bias and decide to approve the drug in the name of reason and decency. Bravo. But what happens when the first patient dies? A media storm ensues, and soon you find yourself out of a job. As a civil servant or politician, you would do well to take the ubiquitous *omission bias* seriously—and even foster it.

Case law shows how engrained such "moral distortion" is in our society. Active euthanasia, even if it is the explicit wish of the dying, is punishable by law, whereas deliberate refusal of lifesaving measures is legal (for example, following so-called DNR orders—do not resuscitate).

Such thinking also explains why parents feel it is perfectly acceptable not to vaccinate their children, even though it discernibly reduces the risk of catching the disease. Of course, there is also a very small risk of getting sick from the vaccine. Overall, however, vaccination makes sense. Vaccination protects not only the children, but society, too. A person who is immune to the disease will never infect others. Objectively, if non-vaccinated children ever contracted one of these sicknesses, we could accuse the parents of actively harming them. But this is exactly the point: Deliberate inaction somehow seems less grave than a comparable action—say, if the parents intentionally infected them.

The *omission bias* lies behind the following delusions: We wait until people shoot themselves in the foot rather than taking aim ourselves. Investors and business journalists are more lenient on companies that develop no new products than they are on those that produce bad ones, even though both roads lead to ruin. Sitting passively on a bunch of miserable shares feels better than actively buying bad ones. Building no emission filter into a coal plant feels superior to removing one for cost reasons. Failing to insulate your house is more acceptable than burning the spared fuel for your own amusement. Neglecting to declare income tax is less immoral than faking tax documents, even though the state loses out either way.

In the previous chapter, we met the *action bias*. Is it the opposite of the *omission bias*? Not quite. The *action bias* causes us to offset a lack of clarity with futile hyperactivity and comes into play when a situation is fuzzy, muddy, or contradictory. The *omission bias*, on the other hand, usually abounds where the situation is intelligible: A future misfortune might be averted with direct action, but this insight doesn't motivate us as much as it should.

The *omission bias* is very difficult to detect—after all, action is more noticeable than inaction. In the 1960s student movements coined a punchy slogan to condemn it: "If you're not part of the solution, you're part of the problem."

45

Don't Blame Me
Self-Serving Bias

Do you ever read annual reports, paying particular attention to the CEO's comments? No? That's a pity, because there you'll find countless examples of this next error, which we all fall for at one time or another. For example, if the company has enjoyed an excellent year, the CEO catalogs his indispensable contributions: his brilliant decisions, tireless efforts, and cultivation of a dynamic corporate culture. However, if the company has had a miserable year, we read about all sorts of other dynamics: the unfortunate exchange rate, governmental interference, the malicious trade practices of the Chinese, various hidden tariffs, subdued consumer confidence, and so on. In short: We attribute success to ourselves and failures to external factors. This is the *self-serving bias*.

Even if you have never heard the expression, you definitely know the *self-serving bias* from high school. If you got an A, you were solely responsible; the top grade reflected your intelligence, hard work, and skill. And if you flunked? The test was clearly unfair.

But grades don't matter to you anymore: Perhaps the stock market has taken their place. There, if you make a profit, you applaud yourself. If your portfolio performs miserably, the blame lies exclusively with "the market" (whatever you imply by this)—or maybe that useless investment adviser. I, too, have periods where I'm a power user of the *self-serving bias*: If my new novel rockets up the best-seller list, I clap myself on the shoulder. Surely this is my best book yet! But if it disappears in the flood of new releases, it is because the readers simply don't recognize good literature when they see it. And if critics slay it, it is clearly a case of jealousy.

To investigate this bias, researchers put together a personality test and afterward allocated the participants' good or bad scores at random. Those who got scored highly found the test thorough and fair; low scorers rated it completely useless. So why do we attribute success to our own skill and ascribe failure to other factors? There are many theories. The simplest explanation is probably this: It feels good. Plus, it doesn't cause any major harm. If it did, evolution would have eliminated it over the past hundred thousand years. But beware: In a modern world with many hidden risks, the *self-serving bias* can quickly lead to catastrophe. Richard Fuld, the self-titled "master of the universe," might well endorse this. He was the almighty CEO of the investment bank Lehman Brothers, until it went bankrupt in 2008. It would not surprise me if he still called himself "master of the universe," blaming government inaction for the bank's collapse.

In SAT tests, students can score between 200 and 800 points. When asked their results a year later, they tend to boost their scores by around 50 points. Interestingly, they are neither

135

lying nor exaggerating; they are simply "enhancing" the result a little—until they start to believe the new score themselves.

In the building where I live, five students share an apartment. I meet them now and again in the elevator, and I decided to ask them separately how often they take out the trash. One said he did it every second time. Another: every third time. Roommate number 3, cursing because his garbage bag had split, reckoned he did it pretty much every time, say 90 percent. Although their answers should have added up to 100 percent, these boys achieved an impressive 320 percent! The five systematically overestimated their roles—and so, are no different from any of us. In married couples, the same thing happens: It's been shown that both men and women overestimate their contribution to the health of the marriage. Each assumes their input is more than 50 percent.

So, how can we dodge the *self-serving bias*? Do you have friends who tell you the truth—no holds barred? If so, consider yourself lucky. If not, do you have at least one enemy? Good. Invite him or her over for coffee and ask for an honest opinion about your strengths and weaknesses. You will be forever grateful you did.

46

Be Careful What You Wish For

Hedonic Treadmill

Suppose one day the phone rings: An excited voice tells you that you have just scooped the lottery jackpot—$10 million! How would you feel? And how long would you feel like that? Another scenario: The phone rings, and you learn that your best friend has passed away. Again, how would you feel, and for how long?

In chapter 40 ("False Prophets: Forecast Illusion"), we examined the miserable accuracy of predictions, for example in the fields of politics, economics, and social events. We concluded that self-appointed experts are of no more use than a random forecast generator. So, moving on to a new area: How well can we predict our feelings? Are we experts on ourselves? Would winning the lottery make us the happiest people alive for years to come? Harvard psychologist Dan Gilbert says no. He has studied lottery winners and discovered that the happiness effect fizzles out after a few months. So, a little while after you receive the big check, you will be as content or as discontent as you were before. He calls this "affective forecasting": our inability to correctly predict our own emotions.

A friend, a banking executive, whose enormous income was beginning to burn a hole in his pocket, decided to build himself a new home away from the city. His dream materialized into a villa with ten rooms, a swimming pool, and an enviable view of the lake and mountains. For the first few weeks, he beamed with delight. But soon the cheerfulness disappeared, and six months later he was unhappier than ever. What happened? As we now know, the happiness effect evaporates after a few months. The villa was no longer his dream. "I come home from work, open the door and . . . nothing. I feel as indifferent about the villa as I did about my one-room student apartment." To make things worse, the poor guy now faced a one-hour commute twice a day. This may sound tolerable, but studies show that commuting by car represents a major source of discontent and stress, and people hardly ever get used to it. In other words, whoever has no innate affinity for commuting will suffer every day—twice a day. Anyhow, the moral of the story is that the dream villa had an overall negative effect on my friend's happiness.

Many others fare no better: People who change or progress in their careers are, in terms of happiness, right back where they started after around three months. The same goes for people who buy the latest Porsche. Science calls this effect the *hedonic treadmill*: We work hard, advance, and are able to afford more and nicer things, and yet this doesn't make us any happier.

So how do negative events affect us—perhaps a spinal cord injury or the loss of a friend? Here, we also overestimate the duration and intensity of future emotions. For example, when a relationship ends, it feels like life will never be the same. The afflicted are completely convinced that they will never again experience joy, but after three or so months, they are back on the dating scene.

Wouldn't it be nice if we knew exactly how happy a new car, career, or relationship would make us? Well, this is doable in part. Use these scientifically rubber-stamped pointers to make better, brighter decisions: (a) Avoid negative things that you cannot grow accustomed to, such as commuting, noise, or chronic stress. (b) Expect only short-term happiness from material things, such as cars, houses, lottery winnings, bonuses, and prizes. (c) Aim for as much free time and autonomy as possible since long-lasting positive effects generally come from what you actively do. Follow your passions even if you must forfeit a portion of your income for them. Invest in friendships. For most people, professional status achieves long-lasting happiness, as long as they don't change peer groups at the same time. In other words, if you ascend to a CEO role and fraternize only with other executives, the effect fizzles out.

47

Do Not Marvel at Your Existence
Self-Selection Bias

Traveling from Philadelphia up to New York, I got stuck in a traffic jam. "Why is it always me?" I groaned. Glancing to the opposite side of the road, I saw carefree southbound drivers racing past with enviable speed. As I spent the next hour crawling forward at a snail's pace, and started to grow restless from braking and accelerating, I asked myself whether I really was especially unlucky. Do I always pick the worst lines at the bank, post office, and grocery store? Or do I just *think* I do?

Suppose that, on this highway, a traffic jam develops 10 percent of the time. The probability that I will get stuck in a jam on a particular day is not greater than the probability that one will occur. However, the likelihood that I will get stuck at a certain point in my journey is greater than 10 percent. The reason: Because I can only crawl forward when in a traffic jam, I spend a disproportionate amount of time in this state. In addition, if the traffic is zooming along, the prospect never crosses my mind. But the moment it arises and I am stuck, I notice it.

The same applies to the lines at bank counters or traffic

lights: Let's say the route between point A and point B has ten traffic lights. On average, one out of the ten will always be red, and the others green. However, you may spend more than 10 percent of your total travel time waiting at a red light. If this doesn't seem right, imagine that you are traveling at near the speed of light. In this case, you would spend 99.99 percent (not 10 percent) of your total journey time waiting and cursing in front of red traffic lights.

Whenever we complain about bad luck, we must be wary of the so-called *self-selection bias*. My male friends often gripe about there being too few women in their companies, and my female friends groan that theirs have too few men. This has nothing to do with bad luck: The grumblers form part of the sample. The probability is high that a man will work in a mostly male industry. Ditto for women. On a grander scale: If you live in a country with a large proportion of men or women (such as China or Russia, respectively), you are likely to form part of the bigger group and accordingly feel hard done by. In elections, it is most probable that you will choose the largest party. In voting, it is most likely that your vote corresponds with the winning majority.

The *self-selection bias* is pervasive. Marketers sometimes stumble into the trap in this way: To analyze how much customers value their newsletter, they send out a questionnaire. Unfortunately, this reaches only one group: current subscribers who are clearly satisfied, have time to respond, and have not canceled their subscriptions. The others make up no part of the sample. Result: The poll is worthless.

Not too long ago, a rather maudlin friend remarked that it bordered on the miraculous that he—yes, he!—ever existed.

A classic victim of the *self-selection bias*. Only someone who is alive can make such an observation. Nonentities generally don't consider their nonexistence for too long. And yet precisely the same delusion forms the basis of at least a dozen philosophers' books, as they marvel year in, year out at the development of language. I'm quite sympathetic to their amazement, but it is simply not justified. If language did not exist, philosophers could not revere it at all—in fact, there would be no philosophers. The miracle of language is tangible only in the environment in which it exists.

Particularly amusing is this recent telephone survey: A company wanted to find out, on average, how many phones (landline and cell) each household owned. When the results were tallied, the firm was amazed that not a single household claimed to have no phone. What a masterpiece.

48

Why Experience Can Damage Your Judgment
Association Bias

Kevin has presented his division's results to the company's board on three occasions. Each time, things have gone perfectly. And, each time, he has worn his green polka-dot boxer shorts. It's official, he thinks: These are my lucky underpants.

The girl in the jewelry store was so stunning that Kevin couldn't help buying the $10,000 engagement ring she showed him. Ten thousand bucks was way over his budget (especially for a second marriage), but for some reason he associated the ring with her and imagined his future wife would be just as dazzling.

Each year, Kevin goes to the doctor for a checkup. Generally, he is told that, for a man of forty-four, he is still in pretty good shape. Only twice has he left the practice with worrying news. Once the problem was his appendix, which was promptly removed. The other time it was a swollen prostate, which, upon further inspection, turned out to be a simple inflammation rather than cancer. Of course, on both occasions, Kevin was beside himself with worry when leaving the clinic—

and coincidentally, both days were extremely hot. Since then, he has always felt uncomfortable on very warm days. If the temperature starts to heat up around one of his checkups, he cancels right away.

Our brain is a connection machine. This is quite practical: If we eat an unknown fruit and feel sick afterward, we avoid it in future, labeling the plant poisonous or at least unpalatable. This is how knowledge comes to be. However, this method also creates false knowledge. Russian scientist Ivan Pavlov was the first to conduct research into this phenomenon. His original goal was to measure salivation in dogs. He used a bell to call the dogs to eat, but soon the ringing sound was enough to make the dogs salivate. The animals' brains linked two functionally unrelated things—the ringing of a bell and the production of saliva.

Pavlov's method works equally well with humans. Advertising creates a link between products and emotions. For this reason, you will never see Coke alongside a frowning face or a wrinkly body. Coke people are young, beautiful, and oh so fun, and they appear in clusters not seen in the real world.

These false connections are the work of the *association bias*, which also influences the quality of our decisions. For example: We often condemn bearers of bad news, since we automatically associate them with the message's content (otherwise known as "shoot-the-messenger syndrome"). Sometimes, CEOs and investors (unconsciously) steer clear of these harbingers, meaning the only news that reaches the upper echelons is positive, thus creating a distorted view of the real situation. If you lead a group of people, and don't want to fall prey to false connections, direct your staff to tell you only the bad news—and fast.

With this, you overcompensate for the shoot-the-messenger syndrome and, believe me, you will still hear enough positive news.

In the days before e-mail and telemarketing, traveling salesmen went door-to-door peddling their wares. One day, a particular salesman, George Foster, stood at a front door. The house transpired to be vacant, and unbeknownst to him, a tiny leak had been filling it with gas for weeks. The bell was also damaged, so when he pressed it, it created a spark and the house exploded. Poor George ended up in the hospital, but fortunately he was soon back on his feet. Unfortunately, his fear of ringing doorbells had become so strong that he couldn't carry out his job for many years. He knew how unlikely a repeat of the incident was, but for all he tried, he just couldn't manage to reverse the (false) emotional connection.

The take-home message from all this is phrased most aptly by Mark Twain: "We should be careful to get out of an experience only the wisdom that is in it—and stop there; lest we be like the cat that sits down on a hot stove-lid. She will never sit down on a hot stove-lid again—and that is well; but also she will never sit down on a cold one anymore."

49

Be Wary When Things Get Off to a Great Start
Beginner's Luck

In the last chapter, we learned about the *association bias*—the tendency to see connections where none exist. For example, regardless of how many big presentations he has nailed while wearing them, Kevin's green polka-dot underpants are no guarantee of success.

We now come to a particularly tricky branch of the *association bias*: creating a (false) link with the past. Casino players know this well; they call it *beginner's luck*. People who are new to a game and lose in the first few rounds are usually clever enough to fold. But whoever strikes lucky tends to keep going. Convinced of their above-average skills, these amateurs increase the stakes—but they soon will get a sobering wake-up call when the probabilities "normalize."

Beginner's luck plays an important role in the economy: Say company A buys smaller companies B, C, and D one after the other. The acquisitions prove a success, and the directors believe they have real skill for acquisitions. Buoyed by this confidence, they now buy a much larger company, E. The integration is a disaster. The merger proves too difficult to handle, the estimated synergies impossible to realize. Objectively speaking, this was foreseeable because in the previous acquisitions everything fell perfectly into place as if guided by a magical hand, so *beginner's luck* blinded them.

The same goes for the stock exchange. Driven by initial success, many investors pumped their life savings into Internet stocks in the late '90s. Some even took out loans to capitalize on the opportunity. However, these investors overlooked one tiny detail: Their amazing profits at the time had nothing to do with their stock-picking abilities. The market was simply on an upward spiral. Even the most clueless investors won big. When the market finally turned downward, many were left facing mountains of dot-com debt.

We witnessed the same delusions during the recent U.S. housing boom. Dentists, lawyers, teachers, and taxi drivers gave up their jobs to "flip" houses—to buy them and resell them right away at higher prices. The first fat profits justified their career changes, but of course these gains had nothing to do with any specific skills. The housing bubble allowed even the most inept amateur brokers to flourish. Many investors became deeply indebted as they flipped even more and even bigger mansions. When the bubble finally burst, many were left with only a string of unsellable properties to their names.

In fact, history has no shortage of *beginner's luck*: I doubt whether Napoleon or Hitler would have dared launch a campaign against the Russians without the previous victories in smaller battles to bolster them.

But how do you tell the difference between *beginner's luck* and the first signs of real talent? There is no clear rule, but these two tips may help: First, if you are much better than others over a long period of time, you can be fairly sure that talent plays a part. (Unfortunately, you can never be 100 percent, though.) Second, the more people competing, the greater the chances are that one of them will repeatedly strike lucky. Perhaps even you. If, among ten competitors, you establish yourself as a market leader over many years, you can clap yourself on the back. That's a sure indication of talent. But if you are top dog among ten million players (i.e., in the financial markets), you shouldn't start visualizing a Buffettesque financial empire just yet; it's extremely likely that you have simply been very fortunate.

Watch and wait before you draw any conclusions. *Beginner's luck* can be devastating, so guard against misconceptions by treating your theories as a scientist would: Try to disprove them. As soon as my first novel, *Thirty-five*, was ready to go, I sent it to a single publisher, where it was promptly accepted. For a moment I felt like a genius, a literary sensation. (The chance that this publisher will take on a manuscript is one in fifteen thousand.) To test my theory, I then sent the manuscript to ten other big publishers. And I got ten rejection letters. My notion was thus disproved, bringing me swiftly back down to earth.

50

Sweet Little Lies

Cognitive Dissonance

A fox crept up to a vine. He gazed longingly at the fat, purple, overripe grapes. He placed his front paws against the trunk of the vine, stretched his neck, and tried to get at the fruit, but it was too high. Irritated, he tried his luck again. He launched himself upward, but his jaw snapped only at fresh air. A third time he leapt with all his might—so powerfully that he landed back down on the ground with a thud. Still not a single leaf had stirred. The fox turned up his nose: "These aren't even ripe yet. Why would I want sour grapes?" Holding his head high, he strode back into the forest.

The Greek poet Aesop created this fable to illustrate one of the most common errors in reasoning. An inconsistency arose when the fox set out to do something and failed to accomplish it. He can resolve this conflict in one of three ways: (a) by somehow getting at the grapes, (b) by admitting that his skills are insufficient, or (c) by reinterpreting what happened retrospectively. The last option is an example of *cognitive dissonance*, or, rather, its resolution.

Suppose you buy a new car. However, you regret your choice soon afterward: The engine sounds like a jet taking off and you just can't get comfortable in the driver's seat. What do you do? Giving the car back would be an admission of error (you don't want that!), and anyway, the dealer probably wouldn't refund all the money. So you tell yourself that a loud engine and awkward seats are great safety features that will prevent you from falling asleep at the wheel. Not so stupid after all, you think, and you are suddenly proud of your sound, practical purchase.

Leon Festinger and James M. Carlsmith of Stanford University once asked their students to carry out an hour of excruciatingly boring tasks. They then divided the subjects into two groups. Each student in group A received a dollar (it was 1959) and instructions to wax lyrical about the work to another student waiting outside—in other words, to lie. The same was asked of the students in group B, with one difference: They were given $20 for the task. Later, the students had to divulge how they really found the monotonous work. Interestingly, those who received only a dollar rated it as significantly more enjoyable and interesting. Why? One measly dollar was not enough for them to lie outright; instead they convinced themselves that the work was not that bad. Just as Aesop's fox reinterpreted the situation, so did they. The students who received more didn't have to justify anything. They had lied and netted $20 for it—a fair deal. They experienced no *cognitive dissonance*.

Suppose you apply for a job and discover you have lost out to another candidate. Instead of admitting that the other person was better suited, you convince yourself that you didn't want the job in the first place; you simply wanted to test your "market value" and see if you could get invited for interview.

I reacted very similarly some time ago when I had to choose between investing in two different stocks. My chosen stock lost much of its value shortly after the purchase, whereas shares in the other stock, the one I hadn't invested in, skyrocketed. I couldn't bring myself to admit my error. Quite the reverse, in fact: I distinctly remember trying to convince a friend that, though the stock was experiencing teething problems, it still had more potential overall. Only *cognitive dissonance* can explain this remarkably irrational reaction. The "potential" would indeed have been even greater if I had postponed the decision to purchase the shares until today. It was that friend who told me the Aesop fable. "You can play the clever fox all you want—but you'll never get the grapes that way."

51

Live Each Day as If It Were Your Last—but Only on Sundays

Hyperbolic Discounting

You know the saying: "Live each day as if it were your last." It features at least three times in every lifestyle magazine and has a slot in every self-help manual's standard repertoire, too. For such a clever line, it makes you none the wiser. Just imagine what would happen if you followed it to the letter: You would no longer brush your teeth, wash your hair, clean the apartment, turn up for work, pay the bills. . . . In no time, you would be broke, sick, and perhaps even behind bars. And yet its meaning is inherently noble: It expresses a deep longing, a desire for immediacy. We place huge value on immediacy—much more than is justifiable. "Enjoy each day to the fullest and don't worry about tomorrow" is simply not a smart way to live.

Would you rather receive $1,000 in a year or $1,100 in a year and a month? Most people will opt for the larger sum in thirteen months—where else will you find a monthly interest rate of 10 percent (or 120 percent per annum!). A wise choice, since the interest will compensate you generously for any risks you face by waiting the extra few weeks.

Second question: Would you prefer $1,000 today cash on the table or $1,100 in a month? If you think like most people, you'll take the $1,000 right away. This is amazing. In both cases, if you hold out for just a month longer, you get $100 more. In the first case, it's simple enough. You figure: "I've already waited twelve months; what's one more?" Not in the second case. The introduction of "now" causes us to make inconsistent decisions. Science calls this phenomenon *hyperbolic discounting*. Put plainly: The closer a reward is, the higher our "emotional interest rate" rises and the more we are willing to give up in exchange for it. The majority of economists have not yet grasped that we respond so subjectively and inconsistently to interest rates. Their models still depend on constant interest rates and are correspondingly questionable.

Hyperbolic discounting, the fact that immediacy magnetizes us, is a remnant of our animal past. Animals will never turn down an instant reward in order to attain more in the future. You can train rats as much as you like; they're never going to give up a piece of cheese today to get two pieces tomorrow. But wait a minute: Don't squirrels manage to gather food and save it for much later? Yes, but that's pure instinct and—verifiably— has nothing to do with impulse control or learning.

And what about children? In the '60s, Walter Mischel conducted a famous experiment on delayed gratification. You can find a wonderful video of this on YouTube by typing in "marshmallow experiment." In it, a group of four-year-olds were each given a marshmallow. They could either eat theirs right away or wait a couple of minutes and receive a second. Amazingly, very few children could wait. Even more amazingly, Mischel found that the capacity for delayed gratification is a reliable indicator of future career success. Patience is indeed a virtue.

The older we get and the more self-control we build up, the more easily we can delay rewards. Instead of twelve months, we happily wait thirteen to take home an additional $100. However, if we are offered an instant reward, the incentive has to be very high for us to postpone the fulfillment. Case in point: the exorbitant interest rates banks charge on credit-card debt and other short-term personal loans, both of which exploit our must-have-now instincts.

In conclusion: Though instantaneous reward is incredibly tempting, *hyperbolic discounting* is still a flaw. The more power we gain over our impulses, the better we can avoid this trap. The less power we have over our impulses—for example, when we are under the influence of alcohol—the more susceptible we are. Viewed from the other side: If you sell consumer products, give customers the option of getting their hands on the items right away. Some people will be willing to pay extra just so they don't have to wait. Amazon makes a bundle from this: A healthy chunk of the next-day delivery surcharge goes directly into its coffers. "Live each day as if it were your last" is a good idea—once a week.

52

Any Lame Excuse

"Because" Justification

Traffic jam on the highway between Los Angeles and San Francisco: surface repairs. I spent thirty minutes slowly battling my way through until the chaos was a distant scene in my rearview mirror. Or so I thought. Half an hour later, I was again bumper to bumper: more maintenance work. Strangely enough, my level of frustration was much lower this time. Why? Reassuringly cheerful signs along the road announced: "We're renovating the highway for you!"

The jam reminded me of an experiment conducted by the Harvard psychologist Ellen Langer in the 1970s. For this, she went into a library and waited at a photocopier until a line had formed. Then she approached the first in line and said: "Excuse me, I have five pages. May I use the Xerox machine?" Her success rate was 60 percent. She repeated the experiment, this time giving a reason: "Excuse me. I have five pages. May I use the Xerox machine because I'm in a rush?" In almost all cases (94 percent), she was allowed to go ahead. This is understandable: If people are in a hurry, you often let them cut in to the front of the line. She tried yet another approach, this time saying: "Excuse me. I have five pages. May I go before you because I have to make some copies?" The result was amazing: Even though the pretext was (a-hem) paper-thin—after all, everyone was standing in line to make copies—she was allowed to pass to the front of the line in almost all cases (93 percent).

When you justify your behavior, you encounter more tolerance and helpfulness. It seems to matter very little if your excuse is good or not. Using the simple validation "because" is sufficient. A sign proclaiming: "We're renovating the highway for you" is completely redundant. What else would a maintenance crew be up to on a highway? If you hadn't noticed before, you realize what is going on once you look out the window. And yet this knowledge reassures and calms you. After all, nothing is more frustrating than being kept in the dark.

Gate A57 at JFK airport, waiting to board: An announcement comes over the loudspeaker: "Attention, passengers. Flight 1234 is delayed by three hours." Wonderful. I walked to the desk to find out why. And came back no more enlightened. I was furious: How dare they leave us waiting in ignorance? Other airlines have the decency to announce: "Flight 5678 is delayed by three hours due to operational reasons." A throwaway reason if ever there was one, but enough to appease passengers.

It seems people are addicted to the word "because"—so much so that we use it even when it's not necessary. If you are a leader, undoubtedly you have witnessed this. If you provide no rallying call, employee motivation dwindles. It simply doesn't make the grade to say that the purpose of your shoe company is to manufacture footwear. No, today, higher purposes and the story behind the story are all-important, such as: "We want our shoes to revolutionize the market" (whatever that means). "Better arch support for a better world!" (whatever that means). Zappo's claims that it is in the happiness business (whatever that means).

If the stock market rises or falls by half a percent, you will never hear the true cause from stock market commentators—that it is white noise, the culmination of an infinite number of market movements. No: People want a palpable reason, and the commentator is happy to select one. Whatever explanation he utters will be meaningless—with frequent blame applied to the pronouncements of Federal Reserve Bank presidents.

If someone asks why you have yet to complete a task, it's best to say: "Because I haven't got around to it yet." It's a pathetic excuse (had you done so, the conversation wouldn't be taking place), but it usually does the trick without the need to scramble for more plausible reasons.

One day I watched my wife carefully separating black laundry from blue. As far as I know, this effort isn't necessary. Both are dark colors, right? Such logic has managed to keep my clothes run-free for many years. "Why do you do that?" I asked. "Because I prefer to wash them separately." For me, a perfectly fine answer.

Never leave home without "because." This unassuming little word greases the wheels of human interaction. Use it unrestrainedly.

53

Decide Better—Decide Less

Decision Fatigue

For weeks, you've been working to the point of exhaustion on this presentation. The PowerPoint slides are polished. Each figure in Excel is indisputable. The pitch is a paradigm of crystal-clear logic. Everything depends on your presentation. If you get the green light from the CEO, you're on your way to a corner office. If the presentation flops, you're on your way to the unemployment office. The CEO's assistant proposes the following times for the presentation: 8:00 a.m., 11:30 a.m., or 6:00 p.m. Which slot do you choose?

The psychologist Roy Baumeister and collaborator Jean Twenge once covered a table with hundreds of inexpensive items—from tennis balls and candles to T-shirts, chewing gum, and Coke cans. He divided his students into two groups. The first group he labeled "deciders," the second, "non-deciders." He told the first group: "I'm going to show you sets containing two random items and each time you have to decide which you prefer. At the end of the experiment I'll give you one item you can take home." They were led to believe that their choices would influence which item they get to keep. To the second group, he said: "Write down what you think about each item, and I'll pick one and give it to you at the end." Immediately thereafter, he asked each student to put their hand in ice cold water and hold it there as long as possible. In psychology, this is a classic method to measure willpower or self-discipline; if you have little or none, you yank your hand back out of the water very quickly. The result: The deciders pulled their hands out of the icy water much sooner than the non-deciders did. The intensive decision making had drained their willpower—an effect confirmed in many other experiments.

Making decisions is exhausting. Anyone who has ever configured a laptop online or researched a long trip—flight, hotels, activities, restaurants, weather—knows this well: After all the comparing, considering, and choosing, you are exhausted. Science calls this *decision fatigue*.

Decision fatigue is perilous: As a consumer, you become more susceptible to advertising messages and impulse buys. As a decision maker, you are more prone to erotic seduction. Willpower is like a battery. After a while it runs out and needs to be recharged. How do you do this? By taking a break, relaxing, and eating something. Willpower plummets to zero if your blood sugar falls too low. IKEA knows this only too well: On the trek through its mazelike display areas and towering warehouse shelves, *decision fatigue* sets in. For this reason, its restaurants are located right in the middle of the stores. The company is willing to sacrifice some of its profit margin so that you can top up your blood sugar on Swedish treats before resuming your hunt for the perfect candlesticks.

Four prisoners in an Israeli jail petitioned the court for early release. Case 1 (scheduled for 8:50 a.m.): an Arab sentenced to thirty months in prison for fraud. Case 2 (scheduled for 1:27 p.m.): a Jew sentenced to sixteen months for assault. Case 3 (scheduled for 3:10 p.m.): a Jew sentenced to sixteen months for assault. Case 4 (scheduled for 4:35 p.m.), an Arab sentenced to thirty months for fraud. How did the judges decide? More significant than the detainees' allegiance or the severity of their crimes was the judges' *decision fatigue*. The judges granted requests 1 and 2 because their blood sugar was still high (from breakfast or lunch). However, they struck out applications 3 and 4 because they could not summon enough energy to risk the consequences of an early release. They took the easy option (the status quo) and the men remained in jail. A study of hundreds of verdicts shows that within a session, the percentage of "courageous" judicial decisions gradually drops from 65 percent to almost zero, and after a recess, returns to 65 percent. So much for the careful deliberations of Lady Justice. But, as long as you have no upcoming trials, all is not lost: You now know when to present your project to the CEO.

54

Would You Wear Hitler's Sweater?

Contagion Bias

Following the collapse of the Carolingian Empire in the ninth century, Europe, especially France, descended into anarchy. Counts, commanders, knights, and other local rulers were perpetually embroiled in battles. The ruthless warriors looted farms, raped women, trampled fields, kidnapped pastors, and set convents alight. Both the Church and the unarmed farmers were powerless against the nobles' savage warmongering.

In the tenth century, a French bishop had an idea. He asked the princes and knights to assemble in a field. Meanwhile, priests, bishops, and abbots gathered all the relics that they could muster from the area and displayed them there. It was a striking sight: bones, blood-soaked rags, bricks, and tiles—anything that had ever come in contact with a saint. The bishop, at that time a person of respect, then called upon the nobles, in the presence of the relics, to renounce unbridled violence and attacks against the unarmed. In order to add weight to his demand, he waved the bloody clothes and holy bones in front of them. The nobles must have had enormous reverence for such symbols: The bishop's unique appeal to their conscience spread throughout Europe, promoting the "Peace and Truce of God." "One should never underestimate the fear of saints in the Middle Ages and of saints' relics," says American historian Philip Daileader.

As an enlightened person, you can only laugh at this silly superstition. But wait: What if I put it to you this way? Would you put on a freshly laundered sweater that Hitler had once worn? Probably not, right? So, it seems that you haven't lost all respect for intangible forces, either. Essentially, this sweater has nothing to do with Hitler anymore. There isn't a single molecule of Hitler's sweat on it. However, the prospect of putting it on still puts you off. It's more than just a matter of respect. Yes, we want to project a "correct" image to our fellow humans and to ourselves, but the thought puts us off even when we are alone and when we convince ourselves that touching this sweater does not endorse Hitler in any way. This emotional reaction is difficult to override. Even those who consider themselves quite rational have a hard time completely banishing the belief in mysterious forces (me included).

Mysterious powers of this kind can't simply be switched off. Paul Rozin and his research colleagues at the University of Pennsylvania asked test subjects to bring in photos of loved ones. These were pinned to the center of targets and the subjects had to shoot darts at them. Riddling a picture with darts does no harm to the person in it but, nevertheless, the subjects' hesitation was palpable. They were much less accurate than a control group that had shot at regular targets beforehand. The test subjects behaved as if a mystic force prevented them from hitting the photos.

The *contagion bias* describes how we are incapable of ignoring the connection we feel to certain items—be they from long ago or only indirectly related (as with the photos). A friend was a longtime war correspondent for the French public television channel France 2. Just as passengers on a Caribbean cruise take home souvenirs from each island—a straw hat or a painted coconut—my friend also collected mementos from her adventures. One of her last missions was to Baghdad in 2003. A few hours after American troops stormed Saddam Hussein's government palace, she crept into the private quarters. In the dining room, she spotted six gold-plated wineglasses and promptly commandeered them. When I attended one of her dinner parties in Paris recently, the gilded goblets had pride of place on the dining table. "Are these from Galeries Lafayette?" one person asked. "No, they are from Saddam Hussein," she said candidly. A horrified guest spat his wine back into the glass and began to splutter uncontrollably. I had to contribute: "You realize how many molecules you've already shared with Saddam, simply by breathing?" I asked. "About a billion per breath." His cough got even worse.

55

Why There Is No Such Thing as an Average War

The Problem with Averages

Suppose you're on a bus with forty-nine other people. At the next stop, the heaviest person in America gets on. Question: By how much has the average weight of the passengers increased? Four percent? Five? Something like that? Suppose the bus stops again, and on gets Bill Gates. This time we are not concerned about weight. Question: By how much has the average *wealth* risen? Four percent? Five? Far from it!

Let's calculate the second example quickly. Suppose each of fifty randomly selected individuals has assets of $54,000. This is the statistical middle value, the median. Then Bill Gates is added to the mix, with his fortune of around $59 billion. The average wealth has just shot up to $1.15 billion, an increase of more than two million percent. A single outlier has radically altered the picture, rendering the term "average" completely meaningless.

"Don't cross a river if it is (*on average*) four feet deep," warns Nassim Taleb, from whom I have the above examples. The river can be very shallow for long stretches—mere inches—but it might transform into a raging torrent that is twenty feet deep in the middle—in which case you could easily drown. Dealing in averages is a risky undertaking because they often mask the underlying distribution—the way the values stack up.

Another example: The average amount of UV rays you are exposed to on a June day is not harmful to your health. But if you were to spend the entire summer in a darkened office, then fly to Barbados and lie in the sun without sunscreen for a week solid, you would have a problem—even though, on average over the summer, you were not getting more UV light than someone who was regularly outside.

All this is quite straightforward and maybe you were aware of it already. For example, you drink one glass of red wine for dinner every evening. That's not a health issue. Many doctors recommend it. But if you drink no alcohol the entire year and on December 31 you gulp 356 glasses, which is equivalent to sixty bottles, you will have a problem, although the *average* over the year is the same.

Here's the update: In a complex world, distribution is becoming more and more irregular. In other words, we will observe the Bill Gates phenomenon in ever more domains. How many visits does an average website get? The answer is: There are no average websites. A handful of sites (such as the *New York Times*, Facebook, or Google) garner the majority of visits, and countless other pages draw comparatively few. In such cases, mathematicians speak of the so-called power law. Take cities. There is one city on this planet with a population of more than thirty million: Tokyo. There are eleven cities with a population of between twenty and thirty million. There are fifteen cities with a population of between ten and twenty million. There are forty-eight cities between five and ten million inhabitants. And thousands (!) between one and five million. That's a power law. A few extremes dominate the distribution, and the concept of average is rendered worthless.

What is the average size of a company? What is the average population of a city? What is an average war (in terms of deaths or duration)? What is the average daily fluctuation in the Dow Jones? What is the average cost overrun of construction projects? How many copies does an average book sell? What is the average amount of damage a hurricane wreaks? What is a banker's average bonus? What is the average success of a marketing campaign? How many downloads does an average iPhone app get? How much money does an average actor earn? Of course you can calculate the answers, but it would be a waste of time. These seemingly routine scenarios are subject to the power law.

To use just the final example: A handful of actors take home more than $10 million per year, while thousands and thousands live on the breadline. Would you advise your son or daughter to get into acting since the average wage is pretty decent? Hopefully not—wrong reason.

In conclusion: If someone uses the word "average," think twice. Try to work out the underlying distribution. If a single anomaly has almost no influence on the set, the concept is still worthwhile. However, when extreme cases dominate (such as the Bill Gates phenomenon), we should discount the term "average." We should all take stock from novelist William Gibson: "The future is already here—it's just not very evenly distributed."

56

How Bonuses Destroy Motivation
Motivation Crowding

A few months ago, a friend from Connecticut decided to move to New York City. This man had a fabulous collection of antiques, such as exquisite old books and handblown Murano glasses from generations ago. I knew how attached he was to them, and how anxious he would be handing them over to a moving company, so the last time I visited, I offered to carry the most fragile items with me when I returned to the city. Two weeks later I got a thank-you letter. Enclosed was a $50 bill.

For years, Switzerland has been considering where to store its radioactive waste. The authorities considered a few different locations for the underground repository, including the village of Wolfenschiessen in the center of the country. Economist Bruno Frey and his fellow researchers at the University of Zurich traveled there and recorded people's opinions at a community meeting. Surprisingly, 50.8 percent were in favor of the proposal. Their positive response can be attributed to several factors: national pride, common decency, social obligation, the prospect of new jobs, and so on. The team carried out the survey a second time, but this time they mentioned a hypothetical reward of $5,000 for each townsperson, paid for by Swiss taxpayers, if they were to accept the proposal. What happened? Results plummeted: Only 24.6 percent were willing to endorse the proposal.

Another example is children's day care centers. Day care workers face the same issue the world over: parents collecting their children after closing time. The staff has no choice but to wait. They can hardly put the last remaining children in taxis or leave them on the curb. To discourage parental tardiness, many nurseries introduced fees for lateness. Ironically, studies show that tardiness actually increased. Of course, they could have instituted a draconian penalty of, say, $500 for each hour—as they could have offered $1 million to each citizen of the small Swiss village. But that's beside the point. The point is: Small—surprisingly small—monetary incentives crowd out other types of incentives.

The three stories illustrate one thing: Money does not always motivate. Indeed, in many cases, it does just the opposite. When my friend slipped me that fifty, he undermined my good deed—and also tainted our friendship. The offer of compensation for the nuclear repository was perceived as a bribe and cheapened the community and patriotic spirit. The nursery's late fees transformed its relationship with parents from interpersonal to monetary, and essentially legitimized their lateness.

Science has a name for this phenomenon: *motivation crowding*. When people do something for well-meaning, nonmonetary reasons—out of the goodness of their hearts, so to speak—payments throw a wrench into the works. Financial reward erodes any other motivations.

Suppose you run a nonprofit organization. Logically, the wages you pay are quite modest. Nevertheless, your employees are highly motivated because they believe they are making a difference. If you suddenly introduce a bonus system—let's say a small salary increase for every donation secured—*motivation crowding* will commence. Your team will begin to snub tasks that bring no extra reward. Creativity, company reputation, knowledge transfer—none of this will matter anymore. Soon, all efforts will zoom in on attracting donations.

So who is safe from *motivation crowding*? This tip should help: Do you know any private bankers, insurance agents, or financial auditors who do their jobs out of passion or who believe in a higher mission? I don't. Financial incentives and performance bonuses work well in industries with generally uninspiring jobs—industries where employees aren't proud of the products or the companies and do the work simply because they get a paycheck. On the other hand, if you create a start-up, you would be wise to enlist employee enthusiasm to promote the company's endeavor rather than try to entice employees with juicy bonuses, which you couldn't pay anyway.

One final tip for those of you who have children: Experience shows that young people are not for sale. If you want your kids to do their homework, practice musical instruments, or even mow the lawn once in a while, do not reach for your wallet. Instead, give them a fixed amount of pocket money each week. Otherwise, they will exploit the system and soon refuse to go to bed without recompense.

57

If You Have Nothing to Say, Say Nothing

Twaddle Tendency

When asked why a fifth of Americans were unable to locate their country on a world map, Miss Teen South Carolina, a high school graduate, gave this answer in front of rolling cameras: "I personally believe that U.S. Americans are unable to do so because some people out there in our nation don't have maps, and I believe that our education like such as South Africa and the Iraq everywhere like such as and I believe that they should our education over here in the U.S. should help the U.S., should help South Africa, and should help the Iraq and the Asian countries, so we will be able to build up our future." The video went viral.

Catastrophic, you agree, but you don't waste too much time listening to beauty queens. Okay, how about the following sentence? "There is certainly no necessity that this increasingly reflexive transmission of cultural traditions be associated with subject-centered reason and future-oriented historical consciousness. To the extent that we become aware of the intersubjective constitution of freedom, the possessive-individualist illusion of autonomy as self-ownership disintegrates." Ring any bells? Top German philosopher and sociologist Jürgen Habermas in *Between Facts and Norms*.

Both of these are manifestations of the same phenomenon, the *twaddle tendency*. Here, reams of words are used to disguise intellectual laziness, stupidity, or underdeveloped ideas. Sometimes it works, sometimes not. For the beauty queen, the smoke screen strategy failed spectacularly. For Habermas, it has worked so far. The more eloquent the haze of words, the more easily we fall for them. If used in conjunction with the *authority bias*, such drivel can be especially dangerous.

I myself have fallen for the *twaddle tendency* on many occasions. When I was younger, French philosopher Jacques Derrida fascinated me. I devoured his books, but even after intense reflection I still couldn't understand much. Subsequently his writings took on a mysterious aura, and the whole experience drove me to write my dissertation on philosophy. In retrospect, both were tomes of useless chatter—Derrida and my dissertation. In my ignorance, I had turned myself into a walking, talking smoke machine.

The *twaddle tendency* is especially rife in sport. Breathless interviewers push equally breathless football players to break down the components of the game, when all they want to say is: "We lost the game—it's really that simple." But the presenter has to fill airtime somehow—and seemingly the best method is by jabbering away, and by compelling the athletes and coaches to join in. Jabber disguises ignorance.

This phenomenon has also taken root in the academic spheres. The fewer results a branch of science publishes, the more babble is necessary. Particularly exposed are economists, which we can see in their comments and economic forecasts.

The same is true for commerce on a smaller scale: The worse off a company is, the greater the talk of the CEO. The extra chatter extends to not just a lot of talking, but to hyperactivity also designed to mask the hardship. A laudable exception is the former CEO of General Electric Jack Welch. He once said in an interview: "You would not believe how difficult it is to be simple and clear. People are afraid that they may be seen as a simpleton. In reality, just the opposite is true."

In conclusion: Verbal expression is the mirror of the mind. Clear thoughts become clear statements, whereas ambiguous ideas transform into vacant ramblings. The trouble is that, in many cases, we lack very lucid thoughts. The world is complicated, and it takes a great deal of mental effort to understand even one facet of the whole. Until you experience such an epiphany, it's better to heed Mark Twain: "If you have nothing to say, say nothing." Simplicity is the zenith of a long, arduous journey, not the starting point.

58

How to Increase the Average IQ of Two States
Will Rogers Phenomenon

Let's say you run a small private bank. The bank manages the money of wealthy and mostly retired individuals. Two money managers—A and B—report to you. Money Manager A manages the money of a few ultra-high-net-worth individuals. Money Manager B has rich, but not extravagantly rich, clients to deal with. The board asks you to increase the average pool of money of both A and B—within six months. If you succeed, you receive a handsome bonus. If not, they'll find someone else to do it. Where do you start?

It's quite simple, actually: You take a client with a sizable but not a huge pool of money from A and give it to B instead. In one fell swoop, this brings up A's average managed wealth as well as B's without you having to find a single new client. The only remaining question is: How will you spend your bonus?

Suppose you switch careers and are now in charge of three hedge funds that invest primarily in privately held companies. Fund A has sensational returns, fund B's are mediocre, and fund C's are miserable. You want to prove yourself to the world, so what's your master plan? You know how it works now: You move a few of A's shares to B and C—picking exactly those investments that have been pulling down A's average returns, but which are still profitable enough to fortify B and C. In no time, all three funds look much healthier. And, because the transformation happened in-house, you don't incur a single fee. Of course, the combined value of the trio hasn't risen by a single cent, but people will still pat you on the back.

This effect is called "stage migration" or the *Will Rogers phenomenon*, named after an American comedian from Oklahoma. He is said to have joked that Oklahomans who pack up and move to California raise both states' average IQ. Since we rarely recognize such scenarios, let's drill the *Will Rogers phenomenon* to anchor it in your memory.

One good example is an auto franchise. Let's say you take charge of two small branches in the same town with a total of six salesmen: numbers 1, 2, and 3 in branch A, and numbers 4, 5, and 6 in branch B. On average, salesman number 1 sells one car per week, salesman number 2 sells two cars per week, and so on up to top salesman number 6, who shifts six cars each week. With a little calculation, you know that branch A sells two cars per salesman, whereas branch B is far ahead with an average of five cars per salesman per week. You decide to transfer salesman number 4 to branch A. What happens? Its average sales increase to 2.5 units per person. And branch B? It now consists of only two salesmen, numbers 5 and 6. Its average sales increase to 5.5 per person. Such switcheroo strategies don't change anything overall, but they create an impressive illusion. For this reason, journalists, investors, and board members should be on special alert when they hear of rising averages in countries, companies, departments, cost centers, or product lines.

A particularly deceitful case of the *Will Rogers phenomenon* is found in medicine. Tumors are usually broken down into four stages: The smallest and most treatable ones are classified as stage one; the worst are rated stage four. Their progression gives us the term "stage migration." The survival rate is highest for stage one patients and lowest for stage four patients. Now, every year new procedures are released onto the market and allow for more accurate diagnosis. These new screening techniques reveal minuscule tumors that no doctor had ever noticed before. The result: Patients who were erroneously diagnosed as healthy before are now counted as stage one patients. The addition of relatively healthy people into the stage one group increases the group's average life expectancy. A great medical success? Unfortunately not: mere *stage migration.*

59

If You Have an Enemy, Give Him Information

Information Bias

In his short story *"Del rigor en la ciencia,"* which consists of just a single paragraph, Jorge Luis Borges describes a special country. In this country, the science of cartography is so sophisticated that only the most detailed of maps will do— that is, a map with a scale of 1:1, as large as the country itself. Their citizens soon realize that such a map does not provide any insight, since it merely duplicates what they already know. Borges's map is the extreme case of the *information bias*, the delusion that more information guarantees better decisions.

Searching for a hotel in Miami a little while ago, I drew up a short list of five good offers. Right away, one jumped out at me, but I wanted to make sure I had found the best deal and decided to keep researching. I plowed my way through dozens of customer reviews and blog posts and clicked through countless photos and videos. Two hours later, I could say for sure which the best hotel was: the one I had liked at the start. The mountain of additional information did not lead to a better decision. On the contrary, if time is money, then I might as well have taken up residence at the Four Seasons.

Jonathan Baron from the University of Pennsylvania asked physicians the following question: A patient presents symptoms that indicate with a probability of 80 percent that he is suffering from disease A. If this is not the case, the patient has either disease X or Y. Each of these diseases is equally bad, and each treatment results in similar side effects. As a doctor, what treatment would you suggest? Logically, you would opt for disease A and recommend the relevant therapy. Now suppose there is a diagnostic test that flashes "positive" when disease X is present and "negative" when disease Y is detected. However, if the patient really does have disease A, the test results will be positive in 50 percent of the cases and negative in the other 50 percent. Would you recommend conducting the test? Most doctors said yes, even though the results would be irrelevant. Assuming that the test result is positive, the probability of disease A is still much greater than that of disease X. The additional information contributes nothing of value to the decision.

Doctors are not the only professionals with a penchant for surplus information. Managers and investors are almost addicted to it. How often are studies commissioned one after the other, even though the critical facts are readily available? Additional information not only wastes time and money, it can also put you at a disadvantage. Consider this question: Which city has more inhabitants, San Diego or San Antonio? Gerd Gigerenzer of the Max Planck Institute in Germany put this question to students in the University of Chicago and the University of Munich. Sixty-two percent of Chicago students guessed right: San Diego has more. But, astonishingly, every single German student answered correctly. The reason: All of them had heard of San Diego but not necessarily of San Antonio, so they opted for the more familiar city. For the Chicagoans, however, both cities were household names. They had more information, and it misled them.

Or consider the hundreds of thousands of economists—in service of banks, think tanks, hedge funds, and governments—and all the white papers they have published from 2005 to 2007: The vast library of research reports and mathematical models. The formidable reams of comments. The polished PowerPoint presentations. The terabytes of information on Bloomberg and Reuters news services. The bacchanal dance to worship the god of information. It was all hot air. The financial crisis touched down and upended global markets, rendering the countless forecasts and comments worthless.

Forget trying to amass all the data. Do your best to get by with the bare facts. It will help you make better decisions. Superfluous knowledge is worthless, whether you know it or not. The historian Daniel J. Boorstin put it right: "The greatest obstacle to discovery is not ignorance—it is the illusion of knowledge." And next time you are confronted by a rival, consider killing him—not with kindness but with reams of data and analysis.

60

Hurts So Good

Effort Justification

J ohn, a soldier in the U.S. Army, has just completed his paratrooper course. He waits patiently in line to receive the coveted parachute pin. At last, his superior officer stands in front of him, lines the pin up against his chest, and pounds it in so hard that it pierces John's flesh. Ever since, he opens his top shirt button at every opportunity to showcase the small scar. Decades later, he has thrown away all the memorabilia from his time in the army, except for the tiny pin, which hangs in a specially made frame on his living-room wall.

Mark single-handedly restored a rusty Harley-Davidson. Every weekend and holiday went into getting it up and running; all the while his marriage was approaching breakdown. It was a struggle, but finally Mark's prized possession was road-ready and gleamed in the sunshine. Two years later, Mark desperately needs money. He sells all his possessions—the TV, the car, even his house—but not the bike. Even when a prospect offers double the actual value, Mark does not sell it.

John and Mark are victims of *effort justification*. When you put a lot of energy into a task, you tend to overvalue the result.

Because John had to endure physical pain for the parachute pin, it outshines all his other awards. And since Mark's Harley cost him so many hours—and also nearly his wife—he prizes the bike so highly that he will never sell it.

Effort justification is a special case of "cognitive dissonance." To have a hole punched in your chest for a simple merit badge borders on the absurd. John's brain compensates for this imbalance by overvaluing the pin, hyping it up from something mundane to something semisacred. All of this happens unconsciously and is difficult to prevent.

Groups use *effort justification* to bind members to them—for example, through initiation rites. Gangs and fraternities initiate new members by forcing them to withstand nauseating or vicious tests. Research proves that the harder the "entrance exam" is to pass, the greater the subsequent pride and the value they attach to their membership. MBA schools play with *effort justification* in this way: They work their students day and night without respite, often to the point of exhaustion. Regardless of whether the course work proves useful later on, once the students have the MBAs in the bag, they'll deem the qualification essential for their careers simply because it demanded so much of them.

A mild form of *effort justification* is the so-called IKEA effect. Furniture that we assemble ourselves seems more valuable than any expensive designer piece. The same goes for handknitted socks. To throw away a handcrafted pair, even if they are tatty and outdated, is hard to do. Managers who put weeks of hard work into a strategy proposal will be incapable of appraising it objectively. Designers, copywriters, product developers, or any other professionals who brood over their creations are similarly guilty of this.

In the '50s, instant cake mixes were introduced to the market. A surefire hit, thought the manufacturers. Far from it: Housewives took an instant disliking to them—because they made things too easy. The firms reacted and made the preparation slightly more difficult (beating in an egg yourself). The added effort raised the ladies' sense of achievement and, with it, their appreciation for convenience food.

Now that you know about *effort justification*, you can rate your projects more objectively. Try it out: Whenever you have invested a lot of time and effort into something, stand back and examine the result—*only* the result. The novel you've been tinkering with for five years and that no publisher wants: Perhaps it's not Nobel-worthy after all. The MBA you felt compelled to do: Would you really recommend it? And the woman you've been chasing for years: Is she really better than bachelorette number two who would say yes right away?

61

Why Small Things Loom Large

The Law of Small Numbers

You sit on the corporate board of a retail company with one thousand stores. Half of the stores are in cities, the other half in rural areas. At the behest of the CEO, a consultant conducted a study on shoplifting and is now presenting his findings. Projected onto the wall in front are the names of the one hundred branches with the highest theft rates compared to sales. In bold letters above them is his eye-opening conclusion: "The branches with the highest theft rate are primarily in rural areas." After a moment of silence and disbelief, the CEO is first to speak: "Ladies and gentlemen, the next steps are clear. From now on, we will install additional safety systems in all rural branches. Let's see those hillbillies steal from us then! Do we all agree?"

Hmmm, not completely. You ask the consultant to call up the hundred branches with the lowest theft rates. After some swift sorting, the list appears. Surprise, surprise: The shops with the least amount of shoplifting in relation to sales are also in rural areas! "The location isn't the deciding factor," you begin, smiling somewhat smugly as you gaze around the table at your colleagues. "What counts is the size of the store. In the countryside, the branches tend to be small, meaning a single incident has a much larger influence on the theft rate. Therefore, the rural branches' rates vary greatly—much more than the larger city branches. Ladies and gentlemen, I introduce you to the *law of small numbers*. It has just caught you out."

The *law of small numbers* is not something we understand intuitively. Thus people—especially journalists, managers, and board members—continually fall for it. Let's examine an extreme example. Instead of the theft rate, consider the average weight of employees in a branch. Instead of a thousand stores, we'll take just two: a mega-branch and a mini-branch. The big store has one thousand employees; the small store just two. The average weight in the large branch corresponds roughly to the average weight of the population, say 170 pounds. Regardless of who is hired or fired, it will not change much. Unlike the small store: The store manager's colleague, if rotund or reedy, will affect the average weight tremendously.

Let's go back to the shoplifting problem. We now understand why the smaller a branch is, the more its theft rate will vary—from extremely high to extremely low. No matter how the consultant arranges his spreadsheet, if you list all the theft rates in order of size, small stores will appear at the bottom, large stores will take up the middle, and the top slots? Small stores again. So, the CEO's conclusion was useless, but at least he doesn't need to go overboard on a security system for the small stores.

Suppose you read the following story in the newspaper: "Start-ups employ smarter people. A study commissioned by the National Institute of Unnecessary Research has calculated the average IQ in American companies. The result: Start-ups hire MENSA material." What is your first reaction? Hopefully a raised eyebrow. This is a perfect example of the *law of small numbers*. Start-ups tend to employ fewer people. Therefore the average IQs will fluctuate much more than those of large corporations, giving small (and new) businesses the highest and lowest scores. The National Institute's study has zero significance. It simply confirms the laws of chance.

So, watch out when you hear remarkable statistics about any small entities: businesses, households, cities, data centers, anthills, parishes, schools, and so on. What is being peddled as an astounding finding is, in fact, a humdrum consequence of random distribution. In his latest book, Nobel Prize winner Daniel Kahneman reveals that even experienced scientists succumb to the *law of small numbers*. How reassuring.

62

Handle with Care

Expectations

On January 31, 2006, Google announced its financial results for the final quarter of 2005. Revenue: up 97 percent. Net profit: up 82 percent. A record-breaking quarter. How did the stock market react to these phenomenal figures? In a matter of seconds, shares tumbled 16 percent. Trading had to be interrupted. When it resumed, the stock plunged another 15 percent. Absolute panic. One particularly desperate trader inquired on his blog: "What's the best skyscraper to throw myself off?" What had gone wrong? Wall Street analysts had anticipated even better results, and when those failed to materialized, $20 billion was slashed from the value of the media giant.

Every investor knows it's impossible to forecast financial results accurately. The logical response to a poor prediction would be: "A bad guess, my mistake." But investors don't react that way. In January 2006, when Juniper Networks announced eagerly anticipated earnings per share that were a *tenth* of a cent lower than analysts' forecasts, the share price fell 21 percent and the company's value plunged $2.5 billion. When expectations are fueled in the run-up to an announcement, any disparity gives rise to draconian punishment, regardless of how paltry the gap is.

Many companies bend over backward to meet analysts' predictions. To escape this terror, some began publishing their own estimates, so-called earnings guidance. Not a smart move. Now, the market heeds only these internal forecasts—and studies them much more closely to boot. CFOs are forced to achieve these targets to the cent, and so must draw on all the accounting artifices available.

Fortunately, expectations can also lead to commendable incentives. In 1965, the American psychologist Robert Rosenthal conducted a noteworthy experiment in various schools. Teachers were told of a (fake) new test that could identify students who were on the verge of an intellectual spurt—so-called bloomers. Twenty percent of students were randomly selected and classified as such. Teachers remained under the impression that these were indeed high-potential students. After a year, Rosenthal discovered that these students had developed much higher IQs than other children in a control group. This effect became known as the "Rosenthal effect" (or "Pygmalion effect").

Unlike the CEOs and CFOs who consciously tailor their performance to meet expectations, the teachers' actions were subconscious. Unknowingly, they probably devoted more time to the bloomers and, consequently, the group learned more. The prospect of brilliant students influenced the teachers so much that they ascribed not just better grades but also improved personality traits to the "gifted" students.

But how do we react to personal expectations? This brings

us to the "placebo effect"—pills and therapies that are unlikely to improve health, but do so anyway. The "placebo effect" has been registered in one-third of all patients. But how it works is not well understood. All we know is that expectations alter the biochemistry of the brain and thus the whole body. Accordingly Alzheimer's patients cannot benefit from it: Their condition impairs the area of the brain that deals with expectations.

Expectations are intangible, but their effect is quite real. They have the power to change reality. Can we deprogram them? Is it possible to live a life free from expectations? Unfortunately not. But you can deal with them more cautiously. Raise expectations for yourself and for the people you love. This increases motivation. At the same time, lower expectations for things you cannot control—for example, the stock market. As paradoxical as it sounds: The best way to shield yourself from nasty surprises is to anticipate them.

63

Speed Traps Ahead!

Simple Logic

Three easy questions. Grab a pen quickly and jot down your answers in the margin. First question: In a department store, a Ping-Pong paddle and a plastic ball cost $1.10. If the paddle costs $1 more, how much is the ball? Second question: In a textile factory, five machines take exactly five minutes to make five shirts. How many minutes will it take one hundred machines to produce one hundred shirts? And, the third question: A pond has water lilies growing in it. The flowers multiply quickly, each day doubling the area they take up. If it takes forty-eight days for the pond to be completely covered with water lilies, how many days will it take for it to be half covered? Don't read on until you have written down the answers.

For each of these questions, there is an intuitive answer— and a right one. The quick, intuitive answers come to mind first: ten cents, one hundred minutes, and twenty-four days. But these are all wrong. The solutions are: five cents, five minutes, and forty-seven days. How many did you answer correctly?

Thousands of people have taken this Cognitive Reflection Test (CRT), which professor Shane Frederick developed. So far, students at the Massachusetts Institute of Technology (MIT) in Boston have fared best. On average, they got 2.18 correct answers. Students at Princeton University came in second with an average of 1.63. Far below were students of the University of Michigan who scored an average of 0.83. However, despite these neat rankings, averages are not interesting in this case. More interesting is how those who scored highly differ from the rest.

Here's a hint: Would you prefer a bird in the hand or two in the bush? Frederick discovered that people with low CRT results tend to prefer a bird in the hand. They play it safe. After all, *something* is better than nothing. Those who score at least 2 or higher usually opt for the riskier option. They prefer the gamble. This is especially true for men.

One factor that separates the groups is their ability to control their impulses. In chapter 51 on *hyperbolic discounting*, we covered the seductive power of "now." Frederick put the following question to the participants: "Would you rather have $3,400 now or $3,800 in a month?" In general, people with low CRT scores favor getting the smaller amount sooner. For them, waiting poses a challenge because they are more impulsive. This also applies to purchasing decisions. In contrast, people with high CRT results usually decide to wait the extra few weeks. They muster the willpower to turn down instant gratification—and are rewarded for it later on.

Thinking is more exhausting than sensing: Rational consideration requires more willpower than simply giving in to intuition. In other words, intuitive people tend to scrutinize less. This led Harvard psychologist Amitai Shenhav and his research colleagues to investigate whether people's CRT results correlate with their faith. Americans with a high CRT score (the study was conducted only in the United States) are often atheists, and their convictions have been reinforced over the years. Participants with low CRT results, however, tend to believe in God and "the immortality of the soul," and often have had divine experiences. This makes sense: The more intuitively people make decisions, the less rationally they query religious beliefs.

If you are less than pleased with your CRT score and want to improve it, start by greeting even the simplest logical questions with incredulity. Not everything that seems plausible is true. Reject the easy answers that pop into your head. So, one more try: You are traveling from A to B. On the way there you drive at 100 mph and on the way back, at 50 mph. What was your average speed? 75 mph? Slow down, slow down!

64

How to Expose a Charlatan

Forer Effect

D ear reader, it may surprise you, but I know you personally. This is how I would sum you up: "You have a great need for other people to like and admire you. You have a tendency to be critical of yourself. You have a great deal of unused capacity, which you have not turned to your advantage. While you have some personality weaknesses, you are generally able to compensate for them. Your sexual adjustment has presented problems for you. Disciplined and self-controlled outside, you tend to be worrisome and insecure inside. At times you have serious doubts as to whether you have made the right decision or done the right thing. You prefer a certain amount of change and variety and become dissatisfied when hemmed in by restrictions and limitations. You pride yourself as an independent thinker and do not accept others' statements without satisfactory proof. You have found it unwise to be too frank in revealing yourself to others. At times you are extroverted, affable, and sociable while at other times you are introverted, wary, and reserved. Some of your aspirations tend to be pretty unrealistic. Security is one of your major goals in life."

Do you recognize yourself? On a scale from 1 (poor) to 5 (excellent), how was my assessment?

In 1948, psychologist Bertram Forer crafted this exact passage using astrology columns from various magazines. He then gave it to his students to read, suggesting that each person was getting a personalized assessment. On average, the students rated their characterizations 4.3 out of 5, that is, they gave Forer an accuracy score of 86 percent. The experiment was repeated hundreds of times in the decades that followed with virtually identical results.

Most likely you gave the text a 4 or 5, too. People tend to identify many of their own traits in such universal descriptions. Science labels this tendency the *Forer effect* (or the "Barnum effect"). The *Forer effect* explains why the pseudosciences work so well—astrology, astrotherapy, the study of handwriting, biorhythm analysis, palmistry, tarot card readings, and séances with the dead.

What's behind the *Forer effect*? First, the majority of statements in Forer's passage are so general that they relate to everyone: "Sometimes you seriously doubt your actions." Who doesn't? Second, we tend to accept flattering statements that don't apply to us: "You are proud of your independent thinking." Obviously! Who sees himself or herself as a mindless follower? Third, the so-called *feature-positive effect* plays a part: The text contains no negative statements; it states only what we are, even though the absence of characteristics is an equally important part of a person's makeup. Fourth, the father of all the fallacies, the *confirmation bias*: We accept whatever corresponds to our self-image and unconsciously filter everything else out. What remains is a coherent portrait.

Whatever tricks astrologers and palm readers can turn, consultants and analysts can, too: "The stock has significant growth potential, even in a very competitive environment. The company lacks the necessary impetus to fully realize and implement ideas from the development team. Management is made up of experienced industry professionals; however, hints of bureaucratization are noticeable. A look at the profit and loss statement clearly shows that savings can be made. We advise the company to focus even more closely on emerging economies to secure future market share." Sounds about right, no?

How do you rate the quality of such a guru—for example, an astrologer? Pick twenty people and secretly assign each a number. Have him characterize the people and write his assessments down on cards. To ensure anonymity, participants never find out their numbers. Afterward, each receives a copy of all the cards. Only when the majority of people identify "their" description is there real talent at hand. I am still waiting.

65

Volunteer Work Is for the Birds

Volunteer's Folly

J ack, a photographer, is on the go from Monday to Friday. Commissioned by fashion magazines, he divides his time between Milan, Paris, and New York and is constantly in search of the most beautiful girls, the most original designs, and the perfect light. He is well known on the social circuit, and the money is great: $500 an hour, easy. "That's as much as a commercial lawyer," he brags to his buddies, "and what I have in front of my lens looks a lot better than any banker."

Jack leads an enviable life, but lately he has become more philosophical. It feels as if something has come between him and the fashion world. The selfishness of the industry suddenly repels him. Sometimes he lies in bed, staring at the ceiling, and yearns for more meaningful work. He would like to be selfless once again, to contribute something to the world, no matter how small.

One day his phone rings. It's Patrick, his former classmate and current president of the local bird club: "Next Saturday we're having our annual birdhouse drive. We're looking for volunteers to help us build birdhouses for endangered species. Afterward we'll put them up in the woods. Do you have time? We're meeting at eight o'clock in the morning. We should be done shortly after noon."

What should Jack say if he really is serious about creating a better world? That's right, he should turn down the request. Why? Jack earns $500 an hour. A carpenter, $50. It would be much more sensible to work an extra hour as a photographer and then hire a professional carpenter for six hours to make good-quality birdhouses (which Jack could never hope to accomplish). Taxes aside, he could donate the difference ($200) to the bird club. Doing so, his contribution would go much further than if he grabbed a saw and rolled up his sleeves.

Nevertheless, it is highly likely that Jack will turn up bright and early next Saturday to build birdhouses. Economists call this *volunteer's folly*. It is a popular phenomenon: More than one-fourth of Americans volunteer their time. But what makes it folly? Among other things, if Jack chooses to cobble together a few birdhouses himself, it takes away work from a tradesman. Working a little longer and donating a portion of the earnings is the most effective contribution Jack can make. Hands-on volunteer work would be helpful only if he could make use of his expertise. If the bird club were planning a fund-raising mail campaign and needed a professional photo, Jack could either shoot it himself or work an hour longer to hire another top photographer and donate the remainder.

So now we come to the thorny topic of altruism. Does selflessness exist at all or is it merely a balm to our egos? Although a desire to help the community motivates many volunteers, personal benefits play a big part, such as gaining skills, experience, and contacts. Suddenly we're not acting quite so selflessly. Indeed, many volunteers engage in what might be deemed "personal happiness management," the benefits of which are sometimes far removed from the real cause. Strictly speaking, anyone who profits or feels even the slightest satisfaction from volunteering is not a pure altruist.

So does it mean Jack is a fool if he turns up, hammer in hand, on Saturday morning? Not necessarily. There is one group exempt from *volunteer's folly*: celebrities. If Bono, Kate Winslet, and Mark Zuckerberg pose for photos while making birdhouses, cleaning oil-stained beaches, or digging for earthquake victims, they lend something priceless to the situation: publicity. Therefore, Jack must critically assess whether he is famous enough to make his participation worthwhile. The same applies to you and me: If people don't double-take when they pass you on the street, the best way to contribute is with greenbacks rather than greenhorn labor.

66

Why You Are a Slave to Your Emotions
Affect Heuristic

What do you think of genetically modified wheat? It's a complex issue. You don't want to answer too hastily. A rational approach would be to consider the controversial technology's pros and cons separately. Write down the possible benefits, weight them in terms of importance, and then multiply them by the probability that they will occur. Doing so, you get a list of expected values. Next, do the same with the cons. List all the disadvantages, estimate their potential damage, and multiply them by the likelihood of them happening. The positive sum minus the negative sum equals the net expected value. If it is above zero, you are in favor of genetically modified wheat. If the sum is below zero, you are against it. More than likely you have already heard of this approach. It is called "expected value," and it features in most literature on decision theory. But just as probable is that you've never bothered to carry out such an evaluation. And without a doubt, none of the professors who wrote the textbooks turned to this method to select their spouses.

Truth be told, no one uses this method to make decisions. First of all, we lack enough imagination to list all the possible pros and cons. We are limited by what springs to mind; we can only conjure up what we have seen in our modest experience. It is hard to imagine a storm of the century if you're only thirty years old. Second, calculating small probabilities is impossible because we do not have enough data on rare events. The smaller the probability, the fewer data points we have and the higher the error rate on the exact probability—a vicious effect. Third, our brain is not built for such calculations. They require time and effort—not our preferred state. In our evolutionary past, whoever thought too long and hard vanished inside a predator's jaws. We are the descendants of quick decision makers, and we rely on mental shortcuts called heuristics.

One of the most popular is the *affect heuristic*. An affect is a momentary judgment: something you like or dislike. The word "gunfire" triggers a negative effect. The word "luxury" produces a positive one. This automatic, one-dimensional impulse prevents you from considering risks and benefits to be independent variables, which indeed they are. Instead, the *affect heuristic* puts risks and benefits on the same sensory thread.

Your emotional reactions to issues such as nuclear power, organic vegetables, private schools, or motorbikes determine how you assess their risks and benefits. If you like something, you believe that the risks are smaller and the benefits greater than they actually are. If you don't like something, the opposite is true. Risks and benefits appear to be dependent. Of course, in reality, they are not.

Even more impressive: Suppose you own a Harley-Davidson. If you come across a study that states that driving one is riskier than previously thought, you will subconsciously tweak how you rate the benefits, deeming the experience "an even greater sense of freedom."

But how does an affect—the initial, spontaneous emotion—come to be? Researchers at the University of Michigan flashed one of three images for less than one hundredth of a second in front of participants: a smiling face, an angry face, or a neutral figure. The subjects then had to indicate whether they liked a randomly selected Chinese character or not (the participants didn't speak Chinese). Most preferred symbols that immediately followed the smiling face. Seemingly insignificant factors influence our emotions. Here is another example where an insignificant factor plays a role. Researchers David Hirschleifer and Tyler Shumway tested the relationship between the amount of morning sun and daily market performance in twenty-six major stock exchanges between 1982 and 1997. They found a correlation that reads much like a farmer's adage: If the sun is shining in the morning, the stock market will rise during the day. Not always, but often. Who would have thought that sunshine can move billions. The morning sun obviously has the same effect as a smiley face.

Whether we like it or not, we are puppets of our emotions. We make complex decisions by consulting our feelings, not our thoughts. Against our best intentions, we substitute the question, "What do I think about this?" with "How do I feel about this?" So, smile! Your future depends on it.

67

Be Your Own Heretic

Introspection Illusion

Bruce is in the vitamin business. His father founded the company when supplements were not yet a lifestyle product; a doctor had to prescribe them. When Bruce took over the operation in the early '90s, demand skyrocketed. Bruce seized the opportunity with both hands and took out huge loans to expand production. Today, he is one of the most successful people in the business and president of a national association of vitamin manufactures. Since childhood, hardly a day has passed without him swallowing at least three multivitamins. A journalist once asked him if they do anything. He replied: "I'm sure of it." Do you believe him?

I have another question for you: Take any idea you are 100 percent sure of: Perhaps that gold will rise over the next five years. Perhaps that God exists. Perhaps that your dentist is overcharging you. Whatever the belief, write it down in one sentence. Do you believe yourself?

I bet you consider your conviction more valid than Bruce's, right? Here's why: Yours is an internal observation, whereas

Bruce's is external. Crudely put, you can peek into your own soul, but not into his.

In Bruce's case, you might think: "Come on, it's obviously in his interest to believe that vitamins are beneficial. After all, his wealth and social status depend on the success of the company. He has to maintain a family tradition. All his life he has gulped down pills, so he'll never admit that it was a waste of time." For you, however, it's a different story: You have searched deep inside. You are completely impartial.

But how pure and honest is internal reflection? The Swedish psychologist Petter Johannson allowed test subjects to glimpse two portrait photos of random people and choose which face was more attractive. Then he showed them the preferred photo up close and asked them to describe the most attractive features. However, with a sleight of hand, he switched the pictures. Most participants failed to notice and proceeded to justify, in detail, why they favored the image. The results of the study: Introspection is not reliable. When we soul-search, we contrive the findings.

The belief that reflection leads to truth or accuracy is called the *introspection illusion*. This is more than sophistry. Because we are so confident of our beliefs, we experience three reactions when someone fails to share our views. Response 1: Assumption of ignorance. The other party clearly lacks the necessary information. If he knew what you know, he would be of the same opinion. Political activists think this way: They believe they can win others over through enlightenment. Reaction 2: Assumption of idiocy. The other person has the necessary information, but his mind is underdeveloped. He cannot draw the obvious conclusions. In other words, he's a moron. This reaction is particularly popular with bureaucrats who want to protect "stupid" consumers from themselves. Response 3: Assumption of malice. Your counterpart has the necessary information—he even understands the debate—but he is deliberately confrontational. He has evil intentions. This is how many religious leaders and followers treat disbelievers: If they don't agree, they must be servants of the devil!

In conclusion: Nothing is more convincing than your own beliefs. We believe that introspection unearths genuine self-knowledge. Unfortunately, introspection is, in large part, fabrication posing two dangers: First, the *introspection illusion* creates inaccurate predictions of future mental states. Trust your internal observations too much and too long, and you might be in for a very rude awakening. Second, we believe that our introspections are more reliable than those of others, which creates an illusion of superiority. Remedy: Be all the more critical with yourself. Regard your internal observations with the same skepticism as claims from some random person. Become your own toughest critic.

68

Why You Should Set Fire to Your Ships
Inability to Close Doors

Next to my bed, two dozen books are stacked high. I have dipped in and out of all of them but am unable to part with even one. I know that sporadic reading won't help me achieve any real insights, despite the many hours I put in, and that I should really devote myself to one book at a time. So why am I still juggling all twenty-four?

I know a man who is dating three women. He is in love with all three and can imagine starting a family with any of them. However, he simply doesn't have the heart to choose just one because then he would be passing up on the other two for good. If he refrains from deciding, all options remain open. The downside is that no real relationship will develop.

In the third century BC, General Xiang Yu sent his army across the Yangtze River to take on the Qin dynasty. While his troops slept, he ordered all the ships to be set alight. The next day he told them: "You now have a choice: Either you fight to win or you die." By removing the option of retreat, he switched their focus to the only thing that mattered: the battle. Spanish conquistador Cortés used the same motivational trick in the sixteenth century. After landing on the east coast of Mexico, he sank his own ship.

Xiang Yu and Cortés are exceptions. We mere mortals do everything we can to keep open the maximum number of options. Psychology professors Dan Ariely and Jiwoong Shin demonstrated the strength of this instinct using a computer game. Players started with one hundred points, and on the screen in front of them, three doors appeared—a red one, a blue one, and a green one. Opening a door cost a point, but for every room they entered, they could accrue more points. The players reacted logically: They found the most fruitful room and holed up there for the whole session. Ariely and Shin then changed the rules. If doors were not opened within twelve moves, they started shrinking on the screen and eventually vanished. Players now rushed from door to door to secure access to all potential treasure troves. All this unproductive scrambling meant they scored 15 percent fewer points than in the previous game. The organizers then added another twist: Opening doors now cost three points. The same anxiety kicked in: Players frittered away their points trying to keep all doors open. Even when the subjects learned how many points were hidden in each room, nothing changed. Sacrificing options was a price they were not willing to pay.

Why do we act so irrationally? Because the downside to such behavior is not always apparent. In the financial markets, things are clear: A financial option on a security always costs something. There is no such thing as a free option. In most other realms, however, options seem to be free. But this is an illusion. They also come at a price, but the price tag is often hidden and intangible: Each decision costs mental energy and eats up precious time for thinking and living. CEOs who examine every possible expansion option often choose none in the end. Companies that aim to address all customer segments end up addressing no one. Salespeople who chase every single lead close no deals.

We are obsessed with having as many irons as possible in the fire, ruling nothing out, and being open to everything. However, this can easily destroy success. We must learn to close doors. A business strategy is primarily a statement on what *not* to engage in. Adopt a life strategy similar to a corporate strategy: Write down what *not* to pursue in your life. In other words, make calculated decisions to disregard certain possibilities and when an option shows up, test it against your not-to-pursue list. It will not only keep you from trouble but also save you lots of thinking time. Think hard once and then just consult your list instead of having to make up your mind whenever a new door cracks open. Most doors are not worth entering, even when the handle seems to turn so effortlessly.

69

Disregard the Brand New

Neomania

How will the world look in fifty years? What will your everyday life be like? With which items will you surround yourself?

People who pondered this question fifty years ago had fanciful notions of how "the future" would look: Highways in the skies. Cities that resemble glass worlds. Bullet trains winding between gleaming skyscrapers. We would live in plastic capsules, work in underwater cities, vacation on the moon, and consume everything in pill form. We wouldn't conceive offspring anymore; instead we would choose children from a catalog. Our best friends would be robots, death would be cured, and we would have exchanged our bikes for jet packs long ago.

But hang on a second. Take a look around. You're sitting in a chair, an invention from ancient Egypt. You wear pants, developed about five thousand years ago and adapted by Germanic tribes around 750 BC. The idea behind your leather shoes comes from the last ice age. Your bookshelves are made of wood, one of the oldest building materials in the world. At dinnertime, you use a fork, a well-known "killer app" from Roman times, to shovel chunks of dead animals and plants into your mouths. Nothing has changed.

How will the world look in fifty years? In his latest book, *Antifragile*, Nassim Taleb gives us a clue: Assume that most of the technology that has existed for the past fifty years will serve us for another half century. And assume that recent technology will be passé in a few years' time. Why? Think of these inventions as if they were species: Whatever has held its own throughout centuries of innovation will probably continue to do so in the future, too. Old technology has proven itself; it possesses an inherent logic even if we do not always understand it. If something has endured for epochs, it must be worth its salt. You can take this to heart the next time you are in a strategy meeting. Fifty years into the future will look a lot like today. Of course, you will witness the birth of many flashy gadgets and magic contraptions. But most will be short-lived.

When contemplating the future, we place far too much emphasis on flavor-of-the-month inventions and the latest "killer apps" while underestimating the role of traditional technology. In the 1960s, space travel was all the rage, so we imagined ourselves on school trips to Mars. In the '70s, plastic was in, so we mulled over how we would furnish our see-through houses. Taleb traces this tendency back to the *neomania* pitfall: the mania for all things shiny and new.

In the past, I sympathized with so-called early adopters, the breed of people who cannot survive without the latest iPhone. I thought they were ahead of their time. Now I regard them as irrational and suffering from a kind of sickness: *neomania*. To them, it is of minor importance if an invention provides tangible benefits; novelty matters more.

So don't go out on a limb when forecasting the future. Stanley Kubrick's cult movie *2001: A Space Odyssey* illustrates why you shouldn't. Made in 1968, the movie predicted that, at the turn of the millennium, the United States would have a thousand-strong colony on the moon and that Pan Am would operate the commuter flights there and back. With this fanciful forecast in mind, I suggest this rule of thumb: Whatever has survived for X years will last another X years. Taleb wagers that the "bullshit filter of history" will sort the gimmicks from the game changers. And that's one bet I'm willing to back.

70

Why Propaganda Works

Sleeper Effect

During World War II, every nation produced propaganda movies. These were devised to fill the population, especially soldiers, with enthusiasm for their country and, if necessary, to bolster them to lay down their lives. The United States spent so much money on propaganda that the War Department decided to find out whether the expense was really worth it. A number of studies were carried out to investigate how the movies affected regular soldiers. The result was disappointing: They did not intensify the privates' enthusiasm for war in the slightest.

Was it because they were poorly made? Hardly. Rather, the soldiers were aware that the movies were propaganda, which discredited their message even before they were rolling. Even if the movie argued a point reasonably or managed to stir the audience, it didn't matter; its content was deemed hollow from the outset and dismissed.

Nine weeks later, something unexpected happened. The psychologists measured the soldiers' attitudes a second time. The result: Whoever had seen the movie expressed much more support for the war than those who had not viewed it. Apparently, propaganda did work after all!

The scientists were baffled, especially since they knew that an argument's persuasiveness decreased over time. It has a half-life like a radioactive substance. Surely you have experienced this yourself: Let's say you read an article on the benefits of gene therapy. Immediately after reading it you are a zealous convert, but after a few weeks, you don't really remember why. More time passes until finally only a tiny fraction of enthusiasm remains.

Amazingly, just the opposite is true for propaganda. If it strikes a chord with someone, this influence will only increase over time. Why? Psychologist Carl Hovland, who led the study for the War Department, named this phenomenon the *sleeper effect*. To date, the best explanation is that, in our memories, the *source* of the argument fades faster than the argument. In other words, your brain quickly forgets where the information came from (e.g., from the Department of Propaganda). Meanwhile, the message itself (i.e., war is necessary and noble) fades only slowly or even endures. Therefore, any knowledge that stems from an untrustworthy source gains credibility over time. The discrediting force melts away faster than the message does.

In the United States, elections increasingly revolve around nasty advertisements, in which candidates seek to tarnish one another's record or reputation. However, by law, each political ad must disclose its sponsor at the end so that it is clearly distinguishable as an electioneering message. However, countless studies show that the *sleeper effect* does its job here, too, especially among undecided voters. The messenger fades from memory; the ugly accusations persevere.

I've often wondered why advertising works at all. Any logical person must recognize ads for what they are, and suitably categorize and disqualify them. But even you as a discerning and intelligent reader won't always succeed at this. It's quite possible that, after a few weeks, you won't remember if you picked up certain information from a well-researched article or from a tacky advertorial.

How can you thwart the *sleeper effect*? First, don't accept any unsolicited advice, even if it seems well meant. Doing so, you protect yourself to a certain degree from manipulation. Second, avoid ad-contaminated sources like the plague. How fortunate we are that books are (still) ad-free! Third, try to remember the source of every argument you encounter. Whose opinions are these? And why do they think that way? Probe the issue like an investigator would: *Cui bono*? Who benefits? Admittedly, this is a lot of work and will slow down your decision making. But it will also refine it.

71

Why It's Never Just a Two-Horse Race

Alternative Blindness

You leaf through a brochure that gushes about the benefits of the university's MBA degree. Your gaze sweeps over photos of the ivy-covered campus and the ultramodern sports facilities. Sprinkled throughout are images of smiling students from various ethnic backgrounds with an emphasis on young women, young Chinese, and young Indian go-getters. On the last page you come to an overview that illustrates the financial value of an MBA. The $100,000 fee is easily offset by the statistical extra income that graduates earn before they retire: $400,000—*after* taxes. Who wouldn't want to be up $300,000? It's a no-brainer.

Wrong. Such an argument hides not one but four fallacies. First, we have the *swimmer's body illusion*: MBA programs attract career-oriented people who will probably earn above-average salaries at some stage of their careers, even without the extra qualification of an MBA. The second fallacy: An MBA takes two years. During this time you can expect a loss of earnings—say, $100,000. So in fact the MBA costs $200,000, not $100,000. That amount, if invested well, could easily exceed the additional income that the brochure promises. Third, to estimate earnings that are more than thirty years away is idiotic. Who knows what will happen over the next three decades? Finally, other alternatives exist. You are not stuck between "do an MBA" and "don't do an MBA." Perhaps you can find a different program that costs significantly less and also represents a shot in the arm for your career. This fourth misconception interests me the most. Let's call it *alternative blindness*: We systematically forget to compare an existing offer with the next-best alternative.

Here's an example from the world of finance: Suppose you have a little money in your savings account, and you ask your investment broker for advice. He proposes a bond that will earn you 5 percent interest. "That's much better than the 1 percent you get with your savings account," he points out. Does it make sense to buy the bond? We don't know. It's wrong to consider just these two options. To assess your options properly, you would have to compare the bond with all other investment options and then select the best. This is how top investor Warren Buffett does things: "Each deal we measure against the second-best deal that is available at any given time—even if it means doing more of what we are already doing."

Unlike Warren Buffett, politicians often fall victim to *alternative blindness*. Let's say your city is planning to build a sports arena on a vacant plot of land. Supporters argue that such an arena would benefit the population much more that an empty lot—both emotionally and financially. But this comparison is wrong. They should compare the construction of the sports arena with all other ideas that become impossible due to its construction—for example, building a school, a performing arts center, a hospital, or an incinerator. They could also sell the land and invest the proceeds or reduce the city's debt.

And you? Do you often overlook the alternatives? Let's say your doctor discovers a tumor that will kill you in five years. He proposes a complicated operation that, if successful, removes the tumor completely. However, this procedure is highly risky, with a survival rate of just 50 percent. How do you decide? You weigh up your choices: Certain death in five years or a 50 percent chance of dying next week. *Alternative blindness!* Perhaps there is a variant of the invasive surgery that your hospital doesn't offer but a hospital across town does. This invasive surgery might not remove the tumor altogether, just slow its growth, but is much safer and gives you an extra ten years. And who knows, maybe during these ten years a more sophisticated therapy for eradicating tumors will be made available.

The bottom line: If you have trouble making a decision, remember that the choices are broader than "no surgery" or "highly risky surgery." Forget about the rock and the hard place, and open your eyes to the other, superior alternatives.

72

Why We Take Aim at Young Guns
Social Comparison Bias

A s one of my books reached number one on the best-seller list, my publisher asked me for a favor. An acquaintance's title was on the verge of entering the top 10 list, and the publisher was convinced that a testimonial from me would give it the necessary push.

It always amazes me that these little testimonials work at all. Everyone knows that only favorable comments end up on a book's jacket. (The book you hold in your hands is no exception.) A rational reader should ignore the praise or at least consider it alongside the criticism, which is always available, albeit in different places. Nevertheless, I've written plenty of testimonials for other books, but they were never for rival titles. I hesitated: Wouldn't writing a blurb be cutting off my nose to spite my face? Why should I help someone who might soon vie for the top slot? As I pondered the question, I realized *social comparison bias* had kicked in—that is, the tendency to withhold assistance to people who might outdo you, even if you look like a fool in the long run.

Book testimonials are a harmless example of the *social comparison bias*. However, the phenomenon has reached toxic levels in academia. Every scientist's goal is to publish as many articles as possible in the most prestigious scientific journals. Over time, you make a name for yourself, and soon editors ask you to assess other scientists' submissions. In the end, often just two or three experts decide what gets published in a particular field. Taking this into account, what happens if a young researcher sends in an earth-shattering paper that turns the entire department on its head and threatens to knock them off their thrones? They will be especially rigorous when evaluating the article. That's *social comparison bias* hard at work.

The psychologist Stephen Garcia and his fellow researchers describe the case of a Nobel laureate who prevented a promising young colleague from applying for a job at "his" university. This may seem judicious in the short term, but in the long run it is counterproductive. What happens when that young prodigy joins another research group and applies his acumen there—most likely depriving the old institution of maintaining its world-class status? Garcia suggests that *social comparison bias* may well be the reason why hardly any research groups remain at the top for many years in succession.

The *social comparison bias* is also a cause for concern with start-up companies. Guy Kawasaki was "chief evangelist" at Apple for four years. Today he is a venture capitalist and advises entrepreneurs. Kawasaki says: "A-players hire people even better than themselves. It's clear, though, that B-players hire C-players so they can feel superior to them, and C-players hire D-players. If you start hiring B-players, expect what Steve [Jobs] called 'the bozo explosion' to happen in your organization." In other words, start hiring B-players and you end up with Z-players. Recommendation: Hire people who are better than you, otherwise you soon preside over a pack of underdogs. The so-called Duning-Kruger effect applies to such Z-players: The inept are gifted at overlooking the extent of their incompetence. They suffer from illusory superiority, which leads them to make even more thinking errors, thus creating a vicious cycle that erodes the talent pool over time.

While his school was closed due to an outbreak of plague in 1666–67, twenty-five-year-old Isaac Newton showed his professor, Isaac Barrow, what research he was conducting in his spare time. Barrow immediately gave up his job as a professor and became a student of Newton. What a noble gesture. What ethical behavior. When was the last time you heard of a professor vacating his post in favor of a better candidate? And when was the last time you read about a CEO clearing out his desk when he realized that one of his twenty thousand employees could do a better job?

In conclusion: Do you foster individuals more talented than you? Admittedly, in the short term, the preponderance of stars can endanger your status, but in the long run, you can only profit from their contributions. Others will overtake you at some stage anyway. Until then, you should get in the up-and-comers' good books—and learn from them. This is why I wrote the testimonial in the end.

73

Why First Impressions Are Deceiving
Primacy and Recency Effects

Allow me to introduce you to two men, Alan and Ben. Without thinking about it too long, decide whom you prefer. Alan is smart, hardworking, impulsive, critical, stubborn, and jealous. Ben, however, is jealous, stubborn, critical, impulsive, hardworking, and smart. Who would you prefer to get stuck with in an elevator? Most people choose Alan, even though the descriptions are exactly the same. Your brain pays more attention to the first adjectives in the lists, causing you to identify two different personalities. Alan is smart and hardworking. Ben is jealous and stubborn. The first traits outshine the rest. This is called the *primacy effect*.

If it were not for the *primacy effect*, people would refrain from decking out their headquarters with luxuriously appointed entrance halls. Your lawyer would feel happy turning up to meet you in worn-out sneakers rather than beautifully polished designer oxfords.

The *primacy effect* triggers practical errors, too. Nobel laureate Daniel Kahneman describes how he used to grade examination papers at the beginning of his professorship. He did it as most teachers do—in order: student 1 followed by student 2 and so on. This meant that students who answered the first questions flawlessly endeared themselves to him, thus affecting how he graded the remaining parts of their exams. So, Kahneman switched methods and began to grade the individual questions in batches—all the answers to question one, then the answers to question two, and so forth. Thus, he canceled out the *primacy effect*.

Unfortunately, this trick is not always replicable. When recruiting a new employee, for example, you run the risk of hiring the person who makes the best first impression. Ideally, you would set up all the candidates in order and let them answer the same question one after the other.

Suppose you sit on the board of a company. A point of discussion is raised—a topic on which you have not yet passed judgment. The first opinion you hear will be crucial to your overall assessment. The same applies to the other participants, a fact that you can exploit: If you have an opinion, don't hesitate airing it first. This way, you will influence your colleagues more and draw them over to your side. If, however, you are chairing the committee, always ask members' opinions in random order so that no one has an unfair advantage.

The *primacy effect* is not always the culprit; the contrasting "recency effect" matters as well. The more recent the information, the better we remember it. This occurs because our short-term memory file drawer, as it were, contains very little extra space. When a new piece of information gets filed, an older piece of information is discarded to make room.

When does the *primacy effect* supersede the *recency effect*, or vice versa? If you have to make an immediate decision based on a series of "impressions" (such as characteristics, exam answers, etc.), the *primacy effect* weighs heavier. But if the series of impressions was formed some time ago, the *recency effect* dominates. For instance, if you listened to a speech a few weeks ago, you will remember the final point or punch line more clearly than your first impressions.

In conclusion: First and last impressions dominate, meaning the content sandwiched between has only a weak influence. Try to avoid evaluations based on first impressions. They will deceive you, guaranteed, in one way or another. Try to assess all aspects impartially. It's not easy, but there are ways around it. For example, in interviews, I jot down a score every five minutes and calculate the average afterward. This way, I make sure that the "middle" counts just as much as hello and good-bye.

74

Why You Can't Beat Homemade
Not-Invented-Here Syndrome

My cooking skills are quite modest, and my wife knows it. However, every now and then I concoct a dish that could pass for edible. A few weeks ago, I bought some sole. Determined to escape the monotony of familiar sauces, I devised a new one—a daring combination of white wine, pureed pistachio nuts, honey, grated orange peel, and a dash of balsamic vinegar. Upon tasting it, my wife slid her baked sole to the edge of the plate and began to scrape off the sauce, smiling ruefully as she did so. I, on the other hand, didn't think it was bad at all. I explained to her in detail what a bold creation she was missing, but her expression stayed the same.

Two weeks later, we were having sole again. This time my wife did the cooking. She prepared two sauces: The first was her tried-and-true beurre blanc, and the other, a new recipe from a French top chef. The second tasted horrible. Afterward, she confessed that it was not a French recipe at all, but a Swiss one: my masterpiece from two weeks before! She had caught me out. I was guilty of the *not-invented-here syndrome* (*NIH syndrome*), which fools us into thinking anything we create ourselves is unbeatable.

NIH syndrome causes you to fall in love with your own ideas. This is valid not only for fish sauces, but for all kinds of solutions, business ideas, and inventions. Companies tend to rate homegrown ideas as far more important than those from outsiders, even if, objectively, this is not the case. I recently had lunch with the CEO of a company that specializes in software for health insurance firms. He told me how difficult it is to sell his software to potential customers, even though his firm is the market leader in terms of service, security, and functionality. Most insurers are convinced that the best solution is what they have crafted themselves in-house over the past thirty years. Another CEO told me how hard it is to get his staff in the company's headquarters to accept solutions proposed from far-flung subsidiaries.

When people collaborate to solve problems and then evaluate these ideas themselves, *NIH syndrome* will inevitably exert an influence. Thus, it makes sense to split teams into two groups. The first group generates ideas, the second rates them, and vice versa. We tend to rate our own business ideas as more successful than other people's concepts. This self-confidence forms the basis of thriving entrepreneurship but also explains start-ups' frequently miserable returns.

This is how psychologist Dan Ariely measured the *NIH syndrome*. Writing in his blog at the *New York Times*, Ariely asked readers to provide solutions to six issues, such as "How can cities reduce water consumption without limiting it by law?" The readers had to make suggestions and to assess the feasibility of all the ideas proposed. They also had to specify how much of their time and money they would invest in each idea. Finally, they were limited to using a set list of fifty words, ensuring that everyone gave more or less the same answers. Despite this, the majority rated their own responses as more important and applicable than the others, even though the submissions were virtually identical.

On a societal level, *NIH syndrome* has serious consequences. We overlook shrewd ideas simply because they come from other cultures. In Switzerland, where each state or "canton" has certain powers, one tiny canton never approved women's suffrage; it took a federal court ruling in 1990 to change the law—a startling case of *NIH*. Or consider the modern traffic roundabout, with its clear yield requirements, that was designed by British transport engineers in the 1960s and implemented throughout the UK. It took another thirty years full of oblivion and resistance until this obvious traffic decongestant found its way in the United States and continental Europe. Today France alone has more than thirty thousand roundabouts, which the French now probably falsely attribute to the designer of the Place de l'Étoile.

In conclusion: We are drunk on our own ideas. To sober up, take a step back every now and then and examine their quality in hindsight. Which of your ideas from the past ten years were truly outstanding? Exactly.

75

How to Profit from the Implausible

The Black Swan

A ll swans are white." For centuries, this statement was watertight. Every snowy specimen corroborated this. A swan in a different color? Unthinkable. That was until the year 1697, when Willem de Vlamingh saw a black swan for the first time during an expedition to Australia. Since then, black swans have become symbols of the improbable.

You invest money in the stock market. Year in, year out, the Dow Jones rises and falls a little. Gradually, you grow accustomed to this gentle up and down. Then, suddenly, a day like October 19, 1987, comes around and the stock market tumbles 22 percent. With no warning. This event is a *Black Swan*, as described by Nassim Taleb in his book with the same title.

A *Black Swan* is an unthinkable event that massively affects your life, your career, your company, your country. There are positive and negative *Black Swans*. The meteorite that flattens you, Sutter's discovery of gold in California, the collapse of the Soviet Union, the invention of the transistor, the Internet browser, the overthrow of Egyptian dictator Mubarak, or another encounter that upturns your life completely—all are *Black Swans*.

Think what you like of former U.S. secretary of defense Donald Rumsfeld, but at a press conference in 2002, he expressed a philosophical thought with exceptional clarity when he offered this observation: There are things we know ("known facts"), there are things we do not know ("known unknowns"), and there are things we do not know that we do not know ("unknown unknowns").

How big is the universe? Does Iran have nuclear weapons? Does the Internet make us smarter or dumber? These are "known unknowns." With enough effort, we can hope to answer these one day. Unlike the "unknown unknowns." No one foresaw Facebook mania ten years ago. It is a *Black Swan*.

Why are *Black Swans* important? Because, as absurd as it may sound, they are cropping up more and more frequently and they tend to become more consequential. Though we can continue to plan for the future, *Black Swans* often destroy our best-laid plans. Feedback loops and nonlinear influences interact and cause unexpected results. The reason: Our brains are designed to help us hunt and gather. Back in the Stone Age, we hardly ever encountered anything truly extraordinary. The deer we chased was sometimes a bit faster or slower, sometimes a little bit fatter or thinner. Everything revolved around a stable mean.

Today is different. With one breakthrough, you can increase your income by a factor of ten thousand. Just ask Larry Page, Usain Bolt, George Soros, J. K. Rowling, or Bono. Such fortunes did not exist previously; peaks of this size were unknown. Only in the most recent of human history has this been possible—hence our problem with extreme scenarios. Since probabilities cannot fall below zero, and our thought processes are prone to error, you should assume that everything has an above-zero probability.

So, what can be done? Put yourself in situations where you can catch a ride on a positive *Black Swan* (as unlikely as that is). Become an artist, inventor, or entrepreneur with a scalable product. If you sell your time (e.g., as an employee, dentist, or journalist), you are waiting in vain for such a break. But even if you feel compelled to continue as such, avoid surroundings where negative *Black Swans* thrive. This means: Stay out of debt, invest your savings as conservatively as possible, and get used to a modest standard of living—no matter whether your big breakthrough comes or not.

76

Knowledge Is Nontransferable

Domain Dependence

Writing books about clear thinking brings with it many pluses. Business leaders and investors invite me to give talks for good money. (Incidentally, this is in itself poor judgment on their part: books are much cheaper.) At a medical conference, the following happened to me. I was speaking about *base-rate neglect* and illustrated it with a medical example: In a forty-year-old patient, stabbing chest pain (among other things) may indicate heart problems as well as stress. Stress is much more frequent (with a higher base rate), so it is advisable to test the patient for this first. All this is very reasonable, and the doctors understood it intuitively. But when I used an example from economics, most faltered.

The same thing happens when I speak in front of investors. If I illustrate fallacies using financial examples, most catch on immediately. However, if I take instances from biology, many are lost. The conclusion: Insights do not pass well from one field to another. This effect is called *domain dependence*.

In 1990, Harry Markowitz received the Nobel Prize in Economics for his theory of "portfolio selection." It describes the optimum composition of a portfolio, taking into account both risk and return prospects. When it came to Markowitz's own portfolio—how he should allot his savings into stocks and bonds—he simply opted for fifty-fifty distribution: half in shares, the other half in bonds. The Nobel Prize winner was incapable of applying his ingenious process to his own affairs. A blatant case of *domain dependence*: He failed to transfer knowledge from the academic world to the private sphere.

A friend of mine is a hopeless adrenaline junkie, scaling overhanging cliffs with his bare hands, and launching himself off mountains in a wing suit. He explained to me last week why starting a business is dangerous: Bankruptcy can never be ruled out. "Personally, I'd rather be bankrupt than dead," I replied. He didn't appreciate my logic.

As an author, I realize just how difficult it is to transfer skills to a new area. For me, devising plots for my novels and creating characters are a cinch. A blank, empty page doesn't daunt me. It's quite a different story with, say, an empty apartment. When it comes to interior decor, I can stand in the room for hours, hands in my pockets, devoid of one single idea.

Business is teeming with *domain dependence*. A software company recruits a successful consumer-goods salesman. The new position blunts his talents; transferring his sales skills from products to services is exceedingly difficult. Similarly, a presenter who is outstanding in front of small groups may well tank when his audience reaches one hundred people. Or a talented marketing mind may be promoted to CEO and suddenly find that he lacks any strategic creativity.

With the Markowitz example, we saw that the transfer from the professional realm to the private realm is particularly difficult to navigate. I know CEOs who are charismatic leaders in the office and hopeless duds at home. Similarly, it would be a hard task to find a more cigarette-toting profession than the prophets of health themselves, the doctors. Police officers are twice as violent at home as civilians. Literary critics' novels get the poorest reviews. And, almost proverbially, the marriages of couples' therapists are frequently more fragile than those of their clients. Mathematics professor Barry Mazur tells this story: "Some years ago I was trying to decide whether or not I should move from Stanford to Harvard. I had bored my friends silly with endless discussion. Finally, one of them said, 'You're one of our leading decision theorists. Maybe you should make a list of the costs and benefits and try to roughly calculate your expected utility.' Without thinking, I blurted out, 'Come on, Sandy, this is serious.'"

What you master in one area is difficult to transfer to another. Especially daunting is the transfer from academia to real life—from the theoretically sound to the practically possible. Of course, this also counts for this book. It will be difficult to transfer the knowledge from these pages to your daily life. Even for me as the writer, that transition proves to be a tough one. Book smarts don't transfer to street smarts easily.

77

The Myth of Like-Mindedness

False-Consensus Effect

Which do you prefer: music from the '60s or music from the '80s? How do you think the general public would answer this question? Most people tend to extrapolate their preferences onto others. If they love the '60s, they will automatically assume that the majority of their peers do, too. The same goes for '80s aficionados. We frequently overestimate unanimity with others, believing that everyone else thinks and feels exactly like we do. This fallacy is called the *false-consensus effect*.

Stanford psychologist Lee Ross hit upon this in 1977. He fashioned a sandwich board emblazoned with the slogan "Eat at Joe's" and asked randomly selected students to wear it around campus for thirty minutes. They also had to estimate how many other students would put themselves forward for the task. Those who declared themselves willing to wear the sign assumed that the majority (62 percent) would also agree to it. On the other hand, those who politely refused believed that most people (67 percent) would find it too stupid to undertake. In both cases, the students imagined themselves to be in the popular majority.

The *false-consensus effect* thrives in interest groups and political factions that consistently overrate the popularity of their causes. An obvious example is global warming. However critical you consider the issue to be, you probably believe that the majority of people share your opinion. Similarly, if politicians are confident of election, it's not just blind optimism: They cannot help overestimating their popularity.

Artists are even worse off. In 99 percent of new projects, they expect to achieve more success than ever before. A personal example: I was completely convinced that my novel *Massimo Marini* would be a resounding success. It was at least as good as my previous books, I thought, and those had done very well. But the public was of a different opinion and I was proven wrong: *false-consensus effect*.

Of course, the business world is equally prone to such false conclusions. Just because an R & D department is convinced of its product's appeal doesn't mean consumers will think the same way. Companies with tech people in charge are especially affected. Inventors fall in love with their products' sophisticated features and mistakenly believe that these will bowl customers over, too.

The *false-consensus effect* is fascinating for yet another reason. If people do not share our opinions, we categorize them as "abnormal." Ross's experiment also corroborated this: The students who wore the sandwich board considered those who refused to be stuck up and humorless, whereas the other camp saw the sign-wearers as idiots and attention seekers.

Perhaps you remember the fallacy of *social proof*, the notion that an idea is better the more people believe in it. Is the *false-consensus effect* identical? No. *Social proof* is an evolutionary survival strategy. Following the crowd has saved our butts more often in the past hundred thousand years than striking out on our own. With the *false-consensus effect*, no outside influences are involved. Despite this, it still has a social function, which is why evolution didn't eliminate it. Our brain is not built to recognize the truth; instead, its goal is to leave behind as many offspring as possible. Whoever seemed courageous and convincing (thanks to the *false-consensus effect*) created a positive impression, attracted a disproportionate amount of resources, and thus increased their chances of passing on their genes to future generations. Doubters were less sexy.

In conclusion: Assume that your worldview is not borne by the public. More than that: Do not assume that those who think differently are idiots. Before you distrust them, question your own assumptions.

78

You Were Right All Along
Falsification of History

Winston Smith, a frail, brooding, thirty-nine-year-old office employee, works in the Ministry of Truth. His job is to update old newspaper articles and documents so that they agree with new developments. His work is important. Revising the past creates the illusion of infallibility and helps the government secure absolute power.

Such historical misrepresentation, as witnessed in George Orwell's classic *1984*, is alive and well today. It may shock you but a little Winston is scribbling away in your brain, too. Worse still: Whereas in Orwell's novel, he toiled unwillingly and eventually rebelled against the system, in your brain he is working with the utmost efficiency and according to your wishes and goals. He will never rise up against you. He revises your memories so effortlessly—elegantly, even—that you never notice his work. Discreet and reliable, Winston disposes of your old, mistaken views. As they vanish one by one, you start to believe you were right all along.

In 1973, U.S. political scientist Gregory Markus asked three thousand people to share their opinions on controversial political issues, such as the legalization of drugs. Their responses ranged from "fully agree" to "completely disagree." Ten years later, he interviewed them again on the same topics, and also asked what they had replied ten years previously. The result: What they recalled disclosing in 1973 was almost identical to their present-day views—and a far cry from their original responses.

By subconsciously adjusting past views to fit present ones, we avoid any embarrassing proof of our fallibility. It's a clever coping strategy because no matter how tough we are, admitting mistakes is an emotionally difficult task. But this is preposterous. Shouldn't we let out a whoop of joy every time we realize we are wrong? After all, such admissions would ensure we will never make the same mistake twice and have essentially taken a step forward. But we do not see it that way.

So does this mean our brains contain no accurately etched memories? Surely not! After all, you can recall the exact moment when you met your partner as if it were captured in a photo. And you can remember exactly where you were on September 11, 2001, when you learned of the terrorist attack in New York, right? You recall to whom you were speaking and how you felt. Your memories of 9/11 are extraordinarily vivid and detailed. Psychologists call these "flashbulb memories": They feel as incontestable as photographs.

They are not. Flashbulb memories are as flawed as regular recollections. They are the product of reconstruction. Ulric Neisser, one of the pioneers in the field of cognitive science, investigated them: In 1986, the day after the explosion of the *Challenger* space shuttle, he asked students to write essays detailing their reactions. Three years later, he interviewed them again. Less than 7 percent of the new data correlated with the initial submissions. In fact, 50 percent of the recollections were incorrect in two-thirds of the points, and 25 percent failed to match even a single detail. Neisser took one of these conflicting papers and presented it to its owner. Her answer: "I know it's my handwriting, but I couldn't have written this." The question remains: Why do flashbulb memories feel so real? We don't know yet.

It is safe to assume that half of what you remember is wrong. Our memories are riddled with inaccuracies, including the seemingly flawless flashbulb memories. Our faith in them can be harmless—or lethal. Consider the widespread use of eyewitness testimony and police lineups to identify criminals. To trust such accounts without additional investigation is reckless, even if the witnesses are adamant that they would easily recognize the perpetrator again.

79

Why You Identify with Your Football Team
In-Group Out-Group Bias

When I was a child, a typical wintery Sunday looked like this: My family sat in front of the TV watching a ski race. My parents cheered for the Swiss skiers and wanted me to do the same. I didn't understand the fuss. First, why zoom down a mountain on two planks? It makes as little sense as hopping up the mountain on one leg, while juggling three balls and stopping every hundred feet to hurl a log as far possible. Second, how can one-hundredth of a second count as a difference? Common sense would say that if people are that close together, they are equally good skiers. Third, why should I identify with the Swiss skiers? Was I related to any of them? I didn't think so. I didn't even know what they thought or read, and if I lived a few feet over the Swiss border, I would probably (have to) cheer for another team altogether.

This brings us to the question: Does identifying with a group—a sports team, an ethnicity, a company, a state—represent flawed thinking?

Over thousands of years, evolution has shaped every behavioral pattern, including attraction to certain groups. In times past, group membership was vital. Fending for yourself was close to impossible. As people began to form alliances, all had to follow suit. Individuals stood no chance against collectives. Whoever rejected membership or got expelled forfeited their place not only in the group, but also in the gene pool. No wonder we are such social animals—our ancestors were, too.

Psychologists have investigated different group effects. These can be neatly categorized under the term *in-group out-group bias*. First, groups often form based on minor, even trivial, criteria. With sports affiliations a random birthplace suffices, and in business it is where you work. To test this, the British psychologist Henri Tajfel split strangers into groups, tossing a coin to choose who went to which group. He told the members of one group it was because they all liked a particular type of art. The results were impressive: Although (a) they were strangers, (b) they were allocated a group at random, and (c) they were far from art connoisseurs, the group members found each other more agreeable than members of other groups. Second, you perceive people outside your own group to be more similar than they actually are. This is called the "out-group homogeneity bias." Stereotypes and prejudices stem from it. Have you ever noticed that, in science-fiction movies, only the humans have different cultures and the aliens do not? Third, since groups often form on the basis of common values, group members receive a disproportionate amount of support for their own views. This distortion is dangerous, especially in business: It leads to the infamous organizational blindness.

Family members helping one another out is understandable. If you share half your genes with your siblings, you are naturally interested in their well-being. But there is such a thing as "pseudo-kinship." It evokes the same emotions without blood relationship. Such feelings can lead to the most idiotic cognitive error of all: laying down your life for a random group—also known as going to war. It is no coincidence that "motherland" suggests kinship. And it's not by chance that the goal of any military training is to forge soldiers together as "brothers."

In conclusion: Prejudice and aversion are biological responses to anything foreign. Identifying with a group has been a survival strategy for hundreds of thousands of years. Not any longer. Identifying with a group distorts your view of the facts. Should you ever be sent to war, and you don't agree with its goals, desert.

80

The Difference between Risk and Uncertainty
Ambiguity Aversion

Two boxes. Box A contains one hundred balls: fifty red and fifty black. Box B also holds one hundred balls, but you don't know how many are red and how many are black. If you reach into one of the boxes without looking and draw out a red ball, you win $100. Which box will you choose: A or B? The majority will opt for A.

Let's play again, using exactly the same boxes. This time, you win $100 if you draw out a *black* ball. Which box will you go for now? Most likely you'll choose A again. But that's illogical! In the first round, you assumed that B contained fewer red balls (and more black balls), so, rationally, you would have to opt for B this time around.

Don't worry; you're not alone in this error—quite the opposite. This result is known as the "Ellsberg Paradox"—named after Daniel Ellsberg, a former Harvard psychologist. (As a side note, he later leaked the top-secret Pentagon Papers to the press, leading to the downfall of President Nixon.) The Ellsberg Paradox offers empirical proof that we favor known probabilities (box A) over unknown ones (box B).

Thus we come to the topics of risk and uncertainty (or ambiguity), and the difference between them. Risk means that the probabilities are known. Uncertainty means that the probabilities are unknown. On the basis of risk, you can decide whether or not to take a gamble. In the realm of uncertainty, though, it's much harder to make decisions. The terms "risk" and "uncertainty" are as frequently mixed up as "cappuccino" and "latte macchiato"—with much graver consequences. You can make calculations with risk, but not with uncertainty. The three-hundred-year-old science of risk is called statistics. A host of professors deal with it, but not a single textbook exists on the subject of uncertainty. Because of this, we try to squeeze ambiguity into risk categories, but it doesn't really fit. Let's look at two examples: one from medicine (where it works) and one from the economy (where it does not).

There are billions of humans on earth. Our bodies do not differ dramatically. We all reach a similar height (no one will ever be one hundred feet tall) and a similar age (no one will live for ten thousand years—or for only a millisecond). Most of us have two eyes, four heart valves, thirty-two teeth. Another species would consider us to be homogeneous—as similar to one another as we consider mice to be. For this reason, there are many similar diseases and it makes sense to say, for example: "There is a 30 percent risk you will die of cancer." On the other hand, the following assertion is meaningless: "There is a 30 percent chance that the euro will collapse in the next five years." Why? The economy resides in the realm of uncertainty. There are not billions of comparable currencies from whose history we can derive probabilities. The difference between risk and uncertainty also illustrates the difference between life insurance and credit default swaps. A credit default swap is an insurance policy against specific defaults, a particular company's inability to pay. In the first case (life insurance), we are in the calculable domain of risk; in the second (credit default swap), we are dealing with uncertainty. This confusion contributed to the chaos of the financial crisis in 2008. If you hear phrases such as "the *risk* of hyperinflation is x percent" or "the *risk* to our equity position is y," start worrying.

To avoid hasty judgment, you must learn to tolerate ambiguity. This is a difficult task and one that you cannot influence actively. Your amygdala plays a crucial role. This is a nut-sized area in the middle of the brain responsible for processing memory and emotions. Depending on how it is built, you will tolerate uncertainty with greater ease or difficulty. This is evident not least in your political orientation: The more averse you are to uncertainty, the more conservatively you will vote. Your political views have a partial biological underpinning.

Either way, whoever hopes to think clearly must understand the difference between risk and uncertainty. Only in very few areas can we count on clear probabilities: casinos, coin tosses, and probability textbooks. Often we are left with troublesome ambiguity. Learn to take it in stride.

81

Why You Go with the Status Quo

Default Effect

In a restaurant the other day I scanned the wine list in desperation. Irouléguy? Harslevelü? Susumaniello? I'm far from an expert, but I could tell that a sommelier was trying to prove his worldliness with these selections. On the last page, I found redemption: "Our French house wine: Réserve du Patron, Bourgogne," $52. I ordered it right away; it couldn't be that bad, I reasoned.

I've owned an iPhone for several years now. The gadget allows me to customize everything—data usage, app synchronization, phone encryption, even how loud I want the camera shutter to sound. How many of these have I set up so far? You guessed it: not one.

In my defense, I'm not technically challenged. Rather, I'm just another victim of the so-called *default effect*. The default setting is as warm and welcoming as a soft pillow, into which we happily collapse. Just as I tend to stick with the house wine and factory cell-phone settings, most people cling to the standard options. For example, new cars are often advertised in a certain color; in every catalog, video, and ad, you see the new car in the same color, although the car is available in a myriad of colors. The percentage of buyers who select this default color far exceeds the percentage of car buyers who bought this particular color in the past. Many opt for the default.

In their book *Nudge*, economist Richard Thaler and law professor Cass Sunstein illustrate how a government can direct its citizens without unconstitutionally restricting their freedom. The authorities simply need to provide a few options—always including a default choice for indecisive individuals. This is how New Jersey and Pennsylvania presented two car-insurance policies to their inhabitants. The first policy was cheaper but waived certain rights to compensation should an accident take place. New Jersey advertised this as the standard option, and most people were happy to take it. In Pennsylvania, however, the second, more expensive option was touted as the standard and promptly became the bestseller. This outcome is quite remarkable, especially when you consider that both states' drivers cannot differ all that much in what they want covered or in what they want to pay.

Or consider this experiment: There is a shortage of organ donors. Only about 40 percent of people opt for it. Scientists Eric Johnson and Dan Goldstein asked people whether, in the event of death, they wanted to actively opt *out* of organ donation. Making donation the default option increased take-up from 40 percent to more than 80 percent of participants, a huge difference between an opt-in and an opt-out default.

The *default effect* is at work even when no standard option is mentioned. In such cases, we make our past the default setting, thereby prolonging and sanctifying the status quo. People crave what they know. Given the choice of trying something new or sticking to the tried-and-tested option, we tend to be highly conservative, even if a change would be beneficial. My bank, for example, charges an annual fee of $60 for mailing out account statements. I could save myself this amount if I downloaded the statements online. However, though the pricey (and paper-guzzling) service has bothered me for years, I still can't bring myself to get rid of it once and for all.

So where does the "status-quo bias" come from? In addition to sheer convenience, *loss aversion* plays a role. Recall that losses upset us twice as much as similar gains please us. For this reason, tasks such as renegotiating existing contracts prove very difficult. Regardless of whether these are private or professional, each concession you make weighs twice as heavy as any you receive, so such exchanges end up feeling like net losses.

Both the *default effect* and the status-quo bias reveal that we have a strong tendency to cling to the way things are, even if this puts us at a disadvantage. By changing the default setting, you can change human behavior.

"Maybe we live our lives according to some grand hidden default idea," I suggested to a dinner companion, hoping to draw him into a deep philosophical discussion. "Maybe it just needs a little time to develop," he said after trying the Réserve du Patron.

82

Why "Last Chances" Make Us Panic

Fear of Regret

Two stories: Paul owns shares in company A. During the year, he considered selling them and buying shares in company B. In the end, he didn't. Today he knows that if he had done so, he would have been up $1,200. Second story: George had shares in company B. During the year, he sold them and bought shares in company A. Today he also knows that if he had stuck with B, he would have netted an extra $1,200. Who feels more regret?

Regret is the feeling of having made the wrong decision. You wish someone would give you a second chance. When asked who would feel worse, 8 percent of respondents said Paul, whereas 92 percent chose George. Why? Considered objectively, the situations are identical. Both Paul and George were unlucky, picked the wrong stock, and were out of pocket for the exact same amount. The only difference: Paul already possessed the shares in A, whereas George went out and bought them. Paul was passive, George active. Paul embodies the majority—most people leave their money lying where it is for years—and George represents the exception. It seems that whoever does not follow the crowd experiences more regret.

It is not always the one who acts who feels more regret. Sometimes, choosing not to act can constitute an exception. An example: A venerable publishing house stands alone in its refusal to publish trendy e-books. Books are made of paper, asserts the owner, and he will stick by this tradition. Shortly afterward, ten publishers go bankrupt. Nine of them attempted to launch e-book strategies and faltered. The final victim is the conventional paper-only publisher. Who will regret the series of decisions most, and who will gain the most sympathy? Right, the stoic e-grumbler.

Here is an example from Daniel Kahneman's book *Thinking, Fast and Slow*: After every plane crash, we hear the story of one unlucky person who actually wanted to fly a day earlier or later, but for some reason he changed his booking at the last minute. Since he is the exception, we feel more sympathy for him than for the other "normal" passengers who were booked on the ill-fated flight from the outset.

The *fear of regret* can make us behave irrationally. To dodge the terrible feeling in the pits of our stomachs, we tend to act conservatively, so as not to deviate from the crowd too much. No one is immune to this, not even supremely self-confident traders. Statistics show that each year on December 31 (D-day for performance reviews and bonus calculations), they tend to off-load their more exotic stocks and conform to the masses. Similarly, *fear of regret* (and the *endowment effect*) prevents you from throwing away things you no longer require. You are afraid of the remorse you will feel in the unlikely event that you needed those worn-out tennis shoes after all.

The *fear of regret* becomes really irksome when combined with a "last chance" offer. A safari brochure promises "the last chance to see a rhino before the species is extinct." If you never cared about seeing one before today, why would you fly all the way to Tanzania to do so now? It is irrational.

Let's say you have long dreamed of owning a house. Land is becoming scarce. Only a handful of plots with sea views are left. Three remain, then two, and now just one. It's your last chance! This thought racing through your head, you give in and buy the last plot at an exorbitant price. The *fear of regret* tricked you into thinking this was a onetime offer, when in reality, real estate with a lake view will always come on the market. The sale of stunning property isn't going to stop anytime soon. "Last chances" make us panic-stricken, and the *fear of regret* can overwhelm even the most hardheaded deal makers.

83

How Eye-Catching Details Render Us Blind
Salience Effect

Imagine the issue of marijuana has been dominating the media for the past few months. Television programs portray potheads, clandestine growers, and dealers. The tabloid press prints photos of twelve-year-old girls smoking joints. Broadsheets roll out the medical arguments and illuminate the societal, even philosophical aspects of the substance. Marijuana is on everyone's lips. Let's assume for a moment that smoking does not affect driving in any way. Just as anyone can wind up in an accident, a driver with a joint is also involved in a crash every now and then—purely coincidentally.

Kurt is a local journalist. One evening, he happens to drive past the scene of an accident. A car is wrapped around a tree trunk. Since Kurt has a very good relationship with the local police, he learns that they found marijuana in the backseat of the car. He hurries back to the newsroom and writes this headline: "Marijuana Kills Yet Another Motorist."

As stated above, we are assuming that the statistical relationship between marijuana and car accidents is zero. Thus, Kurt's headline is unfounded. He has fallen victim to the *salience effect*. *Salience* refers to a prominent feature, a stand-out attribute, a particularity, something that catches your eye. The *salience effect* ensures that outstanding features receive much more attention than they deserve. Since marijuana is the *salient* feature of this accident, Kurt believes that it is responsible for the crash.

A few years later, Kurt moves into business journalism. One of the largest companies in the world has just announced it is promoting a woman to CEO. This is big news! Kurt snaps open his laptop and begins to write his commentary: The woman in question, he types, got the post simply because she is female. In truth, the promotion probably had nothing to do with gender, especially since men fill most top positions. If it were so important to have women as leaders, other companies would have acted by now. But in this news story, gender is the *salient* feature, and thus it earns undue weight.

Not only journalists fall prey to the *salience effect*. We all do. Two men rob a bank and are arrested shortly after. It transpires that they are Nigerian. Although no ethnic group is responsible for a disproportionate number of bank robberies, this *salient* fact distorts our thinking. Lawless immigrants at it again, we think. If an Armenian commits rape, it is attributed to the "Armenians" rather than other factors that also exist among Americans. Thus, prejudices form. That the vast majority of immigrants live lawful lives is easily forgotten. We always recall the undesirable exceptions—they are particularly *salient*. Therefore, whenever immigrants are involved, it is the striking, negative incidents that come to mind first.

The *salience effect* influences not only how we interpret the past but also how we imagine the future. Daniel Kahneman and his fellow researcher Amos Tversky found that we place unwarranted emphasis on *salient* information when we are forecasting. This explains why investors are more sensitive to sensational news (i.e., the dismissal of a CEO) than they are to less striking information (such as the long-term growth of a company's profits). Even professional analysts cannot always evade the *salience effect*.

In conclusion: *Salient* information has an undue influence on how you think and act. We tend to neglect hidden, slow-to-develop, discreet factors. Do not be blinded by irregularities. A book with an unusual, fire-engine red jacket makes it onto the bestseller list. Your first instinct is to attribute the success of the book to the memorable cover. Don't. Gather enough mental energy to fight against seemingly obvious explanations.

84

Why Money Is Not Naked

House-Money Effect

A windy fall day in the early 1980s. The wet leaves swirled about the sidewalk. Pushing my bike up the hill to school, I noticed a strange leaf at my feet. It was big and rust-brown, and only when I bent down did I realize it was a 500–Swiss franc bill! That was the equivalent of about $250 back then, an absolute fortune for a high school student. The money spent little time in my pocket: I soon bought myself a top-of-the-range bike with disc brakes and Shimano gears, one of the best models around. The funny thing was my old bike worked fine.

Admittedly, I wasn't completely broke back then: I had managed to save up a few hundred francs through mowing grass in the neighborhood. However, it never crossed my mind to spend this hard-earned money on something so unnecessary. The most I treated myself to was a trip to the movies every now and then. It was only upon reflection that I realized how irrational my behavior had been. Money is money, after all. But we don't see it that way. Depending on how we get it, we treat it differently. Money is not naked; it is wrapped in an emotional shroud.

Two questions. You've worked hard for a year. At the end of the twelve months, you have $20,000 more in your account than you had at the beginning. What do you do? (a) Leave it sitting in the bank. (b) Invest it. (c) Use it to make necessary improvements, such as renovating your moldy kitchen or replacing old tires. (d) Treat yourself to a luxury cruise. If you think like most people, you'll opt for A, B, or C.

Second question: You win $20,000 in the lottery. What do you do with it? Choose from A, B, C, or D above. Most people now take C or D. And of course, by doing so, they exhibit flawed thinking. You can count it any way you like; $20,000 is still $20,000.

We witness similar delusions in casinos. A friend places $1,000 on the roulette table—and loses everything. When asked about this, he says: "I didn't really gamble away a thousand dollars. I won all that earlier." "But it's the same amount!" "Not for me," he says, laughing.

We treat money that we win, discover, or inherit much more frivolously than hard-earned cash. The economist Richard Thaler calls this the *house-money effect*. It leads us to take bigger risks and, for this reason, many lottery winners end up worse off after they've cashed in their winnings. That old platitude—win some, lose some—is a feeble attempt to downplay real losses.

Thaler divided his students into two groups. The first group learned they had won $30 and could choose to take part in the following coin toss: If it was tails, they would win $9. If heads, they would lose $9. Seventy percent of students opted to risk it. The second group learned they had won nothing but that they could choose between receiving $30 or taking part in a coin toss in which heads won them $21 and tails secured $39. The second group behaved more conservatively. Only 43 percent were prepared to gamble—even though the expected value for both options was the same: $30.

Marketing strategists recognize the usefulness of the *house-money effect*. Online gambling sites "reward" you with $100 credit when you sign up. Credit card companies offer the same when you fill in the application form. Airlines present you with a few thousand miles when you join their frequent-flier clubs. Phone companies give you free call credit to get you accustomed to making lots of calls. A large part of the coupon craze stems from the *house-money effect*.

In conclusion: Be careful if you win money or if a business gives you something for free. Chances are you will pay it back with interest out of sheer exuberance. It's better to tear the provocative clothes from this seemingly free money. Put it in workmen's gear. Put it in your bank account or back into your own company.

85

Why New Year's Resolutions Don't Work
Procrastination

Afriend, a writer, someone who knows how to capture emotion in sentences—let's call him an artist—writes modest books of about a hundred pages every seven years. His output is the equivalent of two lines of print per day. When asked about his miserable productivity, he says: "Researching is just so much more enjoyable than writing." So he sits at his desk, surfing the Web for hours on end or immersed in the most abstruse books—all in the hope of hitting upon a magnificent, forgotten story. Once he has found suitable inspiration, he convinces himself that there is no point starting until he is in the "right mood." Unfortunately, the right mood is a rare occurrence.

Another friend has tried to quit smoking every day for the past ten years. Each cigarette is his last. And me? My tax returns have been lying on my desk for six months, waiting to be completed. I haven't yet given up hope that they will fill themselves in.

Procrastination is the tendency to delay unpleasant but important acts: the arduous trek to the gym, switching to a cheaper insurance policy, writing thank-you letters. Even New Year's resolutions won't help you here.

Procrastination is idiotic because no project completes itself. We know that these tasks are beneficial, so why do we keep pushing them onto the back burner? Because of the time lapse between sowing and reaping. To bridge it requires a high degree of mental energy, as psychologist Roy Baumeister demonstrated in a clever experiment. He put students in front of an oven in which chocolate cookies were baking. Their delicious scent wafted around the room. He then placed a bowl filled with radishes by the oven and told the students that they could eat as many of these as they wanted, but the cookies were strictly out of bounds. He then left the students alone in the room for thirty minutes. Students in a second group were allowed to eat as many cookies as they wanted. Afterward, both groups had to solve a tough math problem. The students who were forbidden to eat any cookies gave up on the math problem twice as fast as those who were allowed to gorge freely on cookies. The period of self-control had drained their mental energy—or willpower—which they now needed to solve the problem. Willpower is like a battery, at least in the short term. If it is depleted, future challenges will falter.

This is a fundamental insight. Self-control is not available around the clock. It needs time to refuel. The good news: To achieve this, all you need to do is refill your blood sugar and kick back and relax.

Though eating enough and giving yourself breaks is important, the next necessary condition is employing an array of tricks to keep you on the straight and narrow. This includes eliminating distractions. When I write a novel, I turn off my

Internet access. It's just too enticing to go online when I reach a knotty part. The most effective trick, however, is to set deadlines. Psychologist Dan Ariely found that dates stipulated by external authorities—for example, a teacher or the IRS—work best. Self-imposed deadlines will work only if the task is broken down step-by-step, with each part assigned its own due date. For this reason, nebulous New Year's resolutions are doomed to fail.

So get over yourself. *Procrastination* is irrational but human. To fight it, use a combined approach. This is how my neighbor managed to write her doctoral thesis in three months: She rented a tiny room with neither telephone nor Internet connection. She set three dates, one for each part of the paper. She told anyone who would listen about these deadlines and even printed them on the back of her business cards. This way, she transformed personal deadlines into public commitments. At lunchtime and in the evenings, she refueled her batteries by reading fashion magazines and sleeping a lot.

86

Build Your Own Castle

Envy

Three scenarios—which would irk you the most? (a) Your friends' salaries increase. Yours stays the same. (b) Their salaries stay the same. Yours is, too. (c) Their average salaries are cut. Yours is, too. If you answered A, don't worry, that's perfectly normal: You're just another victim of the green-eyed monster.

Here is a Russian tale: A farmer finds a magic lamp. He rubs it, and out of thin air a genie appears who promises to grant him one wish. The farmer thinks about this for a little while. Finally, he says: "My neighbor has a cow and I have none. I hope that his drops dead."

As absurd as it sounds, you can probably identify with the farmer. Admit it: A similar thought must have occurred to you at some point in your life. Imagine your colleague scores a big bonus and you get a gift certificate. You feel *envy*. This creates a chain of irrational behavior: You refuse to help him any longer, sabotage his plans, perhaps even puncture the tires of his Porsche. And you secretly rejoice when he breaks his leg skiing.

Of all the emotions, *envy* is the most idiotic. Why? Because it is relatively easy to switch off. This is in contrast to anger, sadness, or fear. "Envy is the most stupid of vices, for there is no single advantage to be gained from it," writes Balzac. In short, envy is the most sincere type of flattery; other than that, it's a waste of time.

Many things spark *envy*: ownership, status, health, youth, talent, popularity, beauty. It is often confused with jealousy because the physical reactions are identical. The difference: The subject of *envy* is a thing (status, money, health, etc.). The subject of jealousy is the behavior of a third person. *Envy* needs two people. Jealousy, on the other hand, requires three: Peter is jealous of Sam because the beautiful girl next door phones him instead.

Paradoxically, with envy, we direct resentments toward those who are most similar to us in age, career, and residence. We don't envy businesspeople from the century before last. We don't begrudge plants or animals. We don't envy millionaires on the other side of the globe—just those on the other side of the city. As a writer, I don't envy musicians, managers, or dentists, but other writers. As a CEO you envy other, bigger CEOs. As a supermodel you envy more successful supermodels. Aristotle knew this: "Potters envy potters."

This brings us to a classic practical error: Let's say your financial success allows you to move from one of New York's grittier neighborhoods to Manhattan's Upper East Side. In the first few weeks, you enjoy being in the center of everything and how impressed your friends are with your new apartment and address. But soon you realize that apartments of completely different proportions surround you. You have traded in your old peer group for one that is much richer. Things start to bother you that haven't bothered you before. *Envy* and status anxiety are the consequences.

How do you curb *envy*? First, stop comparing yourself to others. Second, find your "circle of competence" and fill it on your own. Create a niche where you are the best. It doesn't matter how small your area of mastery is. The main thing is that you are king of the castle.

Like all emotions, *envy* has its origins in our evolutionary past. If the hominid from the cave next door took a bigger share of the mammoth, it meant less for the loser. *Envy* motivated us to do something about it. Laissez-faire hunter-gatherers disappeared from the gene pool; in extreme cases, they died of starvation, while others feasted. We are the offspring of the envious. But, in today's world, *envy* is no longer vital. If my neighbor buys himself a Porsche, it doesn't mean that he has taken anything from me.

When I find myself suffering pangs of *envy*, my wife reminds me: "It's okay to be envious—but only of the person you aspire to become."

87

Why You Prefer Novels to Statistics

Personification

For eighteen years, the American media was prohibited from showing photographs of fallen soldiers' coffins. In February 2009, Defense Secretary Robert Gates lifted this ban and images flooded onto the Internet. Officially, family members have to give their approval before anything is published, but such a rule is unenforceable. Why was the ban created in the first place? To conceal the true costs of war. We can easily find out the number of casualties, but statistics leave us cold. People, on the other hand, especially dead people, spark an emotional reaction.

Why is this? For eons, groups have been essential to our survival. Thus, over the past hundred thousand years, we have developed an impressive sense of how others think and feel. Science calls this the "theory of mind." Here's an experiment to illustrate it: You are given $100 and must share it with a stranger. You can decide how it is divided up. If the other person is happy with your suggestion, the money will be divided that way. If he or she turns down your offer, you must return the $100 and no one gets anything. How do you split the sum?

It would make sense to offer the stranger very little—maybe just a dollar. After all, it's better than nothing. However, in the 1980s, when economists began experimenting with such "ultimatum games" (the technical term), the subjects behaved very differently: They offered the other party between 30 percent and 50 percent. Anything below 30 percent was considered "unfair." The ultimatum game is one of the clearest manifestations of the "theory of mind": In short, we empathize with the other person.

However, with one tiny change, it is possible to almost eliminate this compassion: Put the players in separate rooms. When people can't see their counterparts—or, indeed, when they have never seen them—it is more difficult to simulate their feelings. The other person becomes an abstraction, and the share they are offered drops, on average, to below 20 percent.

In another experiment, psychologist Paul Slovic asked people for donations. One group was shown a photo of Rokia from Malawi, an emaciated child with pleading eyes. Afterward, people donated an average of $2.83 to the charity (out of $5 they were given to fill out a short survey). The second group was shown statistics about the famine in Malawi, including the fact that more than three million malnourished children were affected. The average donation dropped by 50 percent. This is illogical: You would think that people's generosity would grow if they knew the extent of the disaster. But we do not function like that. Statistics don't stir us; people do.

The media has long known that factual reports and bar charts do not entice readers. Hence the guideline: Give the story a face. If a company features in the news, a picture of the CEO appears alongside (either grinning or grimacing, depending on the market). If a state makes the headlines, the president represents it. If an earthquake takes place, a victim becomes the face of the crisis.

This obsession explains the success of a major cultural invention: the novel. This literary "killer app" projects personal and interpersonal conflicts onto a few individual destinies. A scholar could have written a meaty dissertation about the methods of psychological torture in Puritan New England, but instead, we still read Hawthorne's *The Scarlet Letter*. And the Great Depression? In statistical form, this is just a long series of numbers. As a family drama, in Steinbeck's *The Grapes of Wrath*, it is unforgettable.

In conclusion: Be careful when you encounter human stories. Ask for the facts and the statistical distribution behind them. You can still be moved by the story, but this way, you can put it into the right context. If, however, you seek to move and motivate people for your own ends, make sure your tale is seasoned with names and faces.

88

You Have No Idea What You Are Overlooking
Illusion of Attention

After heavy rains in the south of England, a river in a small village overflowed its banks. The police closed the ford, the shallow part of the river where vehicles cross, and diverted traffic. The crossing stayed closed for two weeks, but each day at least one car drove past the warning sign and into the rushing water. The drivers were so focused on their car's navigation systems that they didn't notice what was right in front of them.

In the 1990s, Harvard psychologists Daniel Simons and Christopher Chabris filmed two teams of students passing basketballs back and forth. One team wore black T-shirts, the other, white. The short clip, "The Monkey Business Illusion," is available on YouTube. (Take a look before reading on.) In the video, viewers are asked to count how many times the players in white T-shirts pass the ball. Both teams move in circles, weaving in and out, passing back and forth. Suddenly, in the middle of the video, something bizarre happens: A student dressed as a gorilla walks into the center of the room, pounds his chest, and promptly disappears again. At the end, you are asked if you noticed anything unusual. Half the viewers shake their heads in astonishment. Gorilla? What gorilla?

The monkey business test is considered one of the most famous experiments in psychology and demonstrates the so-called *illusion of attention*: We are confident that we notice everything that takes place in front of us. But in reality, we often see only what we are focusing on—in this case, the passes made by the team in white. Unexpected, unnoticed interruptions can be as large and conspicuous as a gorilla.

The *illusion of attention* can be precarious, for example, when making a phone call while driving. Most of the time doing so poses no problems. The call does not negatively influence the straightforward task of keeping the car in the middle of the lane and braking when a car in front does. But as soon as an unanticipated event takes place, such as a child running across the street, your attention is too stretched to react in time. Studies show that drivers' reactions are equally slow when using a cell phone as when under the influence of alcohol or drugs. Furthermore, it does not matter whether you hold the phone with one hand, jam it between your shoulder and jaw, or use a hands-free kit: Your responsiveness to unexpected events is still compromised.

Perhaps you know the expression "the elephant in the room." It refers to an obvious subject that nobody wants to discuss, a kind of taboo. In contrast, let us define what "the gorilla in the room" is: a topic that is of the utmost importance and urgency, and that we absolutely need to address, but nobody knows about it.

Take the case of Swissair, a company that was so fixated on expansion that it overlooked its evaporating liquidity and went bankrupt in 2001. Or the mismanagement in the Eastern bloc that led to the fall of the Berlin Wall. Or the risks on banks' books that up until 2007 nobody paid any attention to. Such gorillas stomp around right in front of us—and we barely spot them.

It's not the case that we miss every extraordinary event. The crux of the matter is that whatever we fail to notice remains unheeded. Therefore, we have no idea what we are overlooking. This is exactly why we still cling to the dangerous illusion that we perceive everything of importance.

Purge yourself of the *illusion of attention* every now and then. Confront all possible and seemingly impossible scenarios. What unexpected events might happen? What lurks beside and behind the burning issues? What is no one addressing? Pay attention to silences as much as you respond to noises. Check the periphery, not just the center. Think the unthinkable. Something unusual can be huge; we still may not see it. Being big and distinctive is not enough to be seen. The unusual and huge thing must be expected.

89

Hot Air
Strategic Misrepresentation

Suppose you apply for your dream job. You buff your résumé to a shine. In the job interview, you highlight your achievements and abilities and gloss over weak points and setbacks. When they ask if you could boost sales by 30 percent while cutting costs by 30 percent, you reply in a calm voice: "Consider it done." Even though you are trembling inside and racking your brain about how the hell you are going to pull that off, you do and say whatever is necessary to get the job. You concentrate on wowing the interviewers; the details will follow. You know that if you give even semi-realistic answers, you'll put yourself out of the race.

Imagine you are a journalist and have a great idea for a book. The issue is on everyone's lips. You find a publisher who is willing to pay a nice advance. However, he needs to know your timeline. He removes his glasses and looks at you: "When can I expect the manuscript? Can you have it ready in six months?" You gulp. You've never written a book in under three years. Your answer: "Consider it done." Of course you don't want to lie, but you know that you won't get the advance if you tell the truth. Once the contract is signed and the money is nestling in your bank account, you can always keep the publisher at bay for a while. You're a writer; you're great at making up stories!

The official term for such behavior is *strategic misrepresentation*: the more at stake, the more exaggerated your assertions become. *Strategic misrepresentation* does not work everywhere. If your ophthalmologist promises five times in a row to give you perfect vision, but after each procedure you see worse than before, you will stop taking him seriously at some point. However, when unique attempts are involved, *strategic misrepresentation* is worth a try—in interviews, for example, as we saw above. A single company isn't going to hire you several times. It's either a yes or no.

Most vulnerable to *strategic misrepresentation* are mega-projects, where (a) accountability is diffuse (for example, if the administration that commissioned the project is no longer in power), (b) many businesses are involved, leading to mutual finger-pointing, or (c) the end date is a few years down the road.

No one knows more about large-scale projects than Oxford professor Bent Flyvbjerg. Why are cost and schedule overruns so frequent? Because it is not the best offer overall that wins; it is whichever one looks best on paper. Flyvbjerg calls this "reverse Darwinism": Whoever produces the most hot air will be rewarded with the project. However, is *strategic misrepresentation* simply brazen deceit? Yes and no. Are women who wear makeup frauds? Are men who lease Porsches to signal financial prowess liars? Yes and no. Objectively they are, but the deceit is socially acceptable, so we don't get worked up about it. The same counts for *strategic misrepresentation*.

In many cases, *strategic misrepresentation* is harmless. However, for the things that matter, such as your health or future employees, you must be on your guard. So, if you are dealing with a person (a first-rate candidate, an author, or an ophthalmologist), don't go by what they claim; look at their past performance. When it comes to projects, consider the timeline, benefits, and costs of similar projects, and grill anyone whose proposals are much more optimistic. Ask an accountant to pick apart the plans mercilessly. Add a clause into the contract that stipulates harsh financial penalties for cost and schedule overruns. And, as an added safety measure, have this money transferred to a secure escrow account.

90

Where's the Off Switch?
Overthinking

There was once an intelligent centipede. Sitting on the edge of a table, he looked over and saw a tasty grain of sugar across the room. Clever as he was, he started to weigh up the best route: Which table leg should he crawl down—left or right—and which table leg should he crawl up? The next tasks were to decide which foot should take the first step, in which order the others should follow, and so on. He was adept at mathematics, so he analyzed all the variants and selected the best path. Finally, he took the first step. However, still engrossed in calculation and contemplation, he got tangled up and stopped dead in his tracks to review his plan. In the end, he came no further and starved.

The British Open golf tournament in 1999: French golfer Jean van de Velde played flawlessly until the final hole. With a three-shot lead, he could easily afford a double bogey (two over par) and still win. Child's play! Entry into the big leagues was now only a matter of minutes away. All he needed to do was to play it safe. But as Van de Velde stepped up, beads of sweat began to form on his forehead. He teed off like a beginner. The ball sailed into the bushes, landing almost twenty feet from the hole. He became increasingly nervous. The next shots were no better. He hit the ball into knee-high grass, then into the water. He took off his shoes, waded into the water, and for a minute contemplated shooting from the pond. But he decided to take the penalty. He then shot into the sand. His body movements suddenly resembled those of a novice. Finally, he made it onto the green and—after a seventh attempt—into the hole. Van de Velde lost the British Open and secured a place in sporting history with his now-notorious triple bogey. It was the beginning of the end of his career. (He celebrated an impressive comeback in 2005.)

In the 1980s, *Consumer Reports* asked experienced tasters to sample forty-five different varieties of strawberry jelly. A few years later, psychology professors Timothy Wilson and Jonathan Schooler repeated the experiment with students from the University of Washington. The results were almost identical. Both students and experts preferred the same type. But that was only the first part of Wilson's experiment. He repeated it with a second group of students who, unlike the first group, had to fill in a questionnaire justifying their ratings in detail. The rankings turned out to be completely warped. Some of the best varieties ended up at the bottom of the rankings.

Essentially, if you think too much, you cut off your mind from the wisdom of your feelings. This may sound a little esoteric—and a bit surprising coming from someone like me who strives to rid my thinking of irrationality—but it is not. Emotions form in the brain, just as crystal-clear, rational thoughts do. They are merely a different form of information processing—more primordial, but not necessarily an inferior variant. In fact, sometimes they provide the wiser counsel.

This raises the question: When do you listen to your head and when do you heed your gut? A rule of thumb might be: If it is something to do with practiced activities, such as motor skills (think of the centipede, Van de Velde, or mastering a musical instrument) or questions you've answered a thousand times (think of Warren Buffett's "circle of competence"), it's better not to reflect to the last detail. It undermines your intuitive ability to solve problems. The same applies to decisions that our Stone Age ancestors faced—evaluating what was edible, who would make good friends, whom to trust. For such purposes, we have heuristics, mental shortcuts that are clearly superior to rational thought. With complex matters, though, such as investment decisions, sober reflection is indispensable. Evolution has not equipped us for such considerations, so logic trumps intuition.

91

Why You Take On Too Much

Planning Fallacy

Every morning, you compile a to-do list. How often does it happen that everything is checked off by the end of the day? Always? Every other day? Maybe once a week? If you are like most people, you will achieve this rare state once a month. In other words, you systematically take on too much. More than that: Your plans are absurdly ambitious. Such a thing would be forgivable if you were a planning novice. But you've been compiling to-do lists for years, if not decades. Thus, you know your capabilities inside out and it's unlikely that you overestimate them afresh every day. This is not facetiousness: In other areas, you learn from experience. So why is there no learning curve when it comes to making plans? Even though you realize that most of your previous endeavors were overly optimistic, you believe in all seriousness that, today, the same workload—or more—is eminently doable. Daniel Kahneman calls this the *planning fallacy*.

In their last semesters, students generally have to write theses. The Canadian psychologist Roger Buehler and his research team asked the following of their final-year class: The students had to specify two submission dates: The first was a "realistic" deadline and the second was a "worst-case scenario" date. The result? Only 30 percent of students made the realistic deadlines. On average, the students needed 50 percent more time than planned—and a full seven days more than their worst-case scenario date.

The *planning fallacy* is particularly evident when people work together—in business, science, and politics. Groups overestimate duration and benefits and systematically underestimate costs and risks. The conch-shaped Sydney Opera House was planned in 1957: Completion was due in 1963 at a cost of $7 million. It finally opened its doors in 1973 after $102 million had been pumped in—fourteen times the original estimate!

So why are we not natural-born planners? The first reason: wishful thinking. We want to be successful and achieve everything we take on. Second, we focus too much on the project and overlook outside influences. Unexpected events too often defeat our plans. This is true for daily schedules, too: Your daughter swallows a fish bone. Your car battery gives up the ghost. An offer for a house lands on your desk and must be discussed urgently. There goes the plan. If you planned things even more minutely, would that be a solution? No, step-by-step preparation amplifies the *planning fallacy*. It narrows your focus even more and thus distracts you even more from anticipating the unexpected.

So what can you do? Shift your focus from internal things, such as your own project, to external factors, like similar projects. Look at the base rate and consult the past. If other ventures of the same type lasted three years and devoured $5 million, this will probably apply to your project, too—no matter how carefully you plan. And, most important, shortly before decisions are made, perform a so-called premortem session (literally, "before death"). American psychologist Gary Klein recommends delivering this short speech to the assembled team: "Imagine it is a year from today. We have followed the plan to the letter. The result is a disaster. Take five or ten minutes to write about this disaster." The stories will show you how things might turn out.

92

Those Wielding Hammers See Only Nails
Déformation Professionnelle

A man takes out a loan, starts a company, and goes bankrupt shortly afterward. He falls into a depression and commits suicide.

What do you make of this story? As a business analyst, you want to understand why the business idea did not work: Was he a bad leader? Was the strategy wrong, the market too small, or the competition too large? As a marketer, you imagine the campaigns were poorly organized or that he failed to reach his target audience. If you are a financial expert, you ask whether the loan was the right financial instrument. As a local journalist, you realize the potential of the story: How lucky that he killed himself! As a writer, you think about how the incident could develop into a kind of Greek tragedy. As a banker, you believe an error took place in the loan department. As a socialist, you blame the failure of capitalism. As a religious conservative, you see in this a punishment from God. As a psychiatrist, you recognize low serotonin levels. Which is the "correct" viewpoint?

None of them. "If your only tool is a hammer, all your problems will be nails," said Mark Twain—a quote that sums up the *déformation professionnelle* perfectly. Charlie Munger, Warren Buffett's business partner, named the effect the "man with the hammer tendency" after Twain: "But that's a perfectly disastrous way to think and a perfectly disastrous way to operate in the world. So you've got to have multiple models. And the models have to come from multiple disciplines—because all the wisdom of the world is not to be found in one little academic department."

Here are a few examples of *déformation professionnelle*: Surgeons want to solve almost every medical problem with a scalpel, even if their patients could be treated with less invasive methods. Armies think of military solutions first. Engineers, structural. Trend gurus see trends in everything (incidentally, this is one of the most idiotic ways to view the world). In short: If you ask people the crux of a particular problem, they usually link it to their own areas of expertise.

So what's wrong with that? It's good if, say, a tailor sticks to what he knows. The *déformation professionnelle* becomes hazardous when people apply their specialized processes in areas where they don't belong. Surely you've come across some of these: Teachers who scold their friends like students. New mothers who begin to treat their husbands like children. Or consider the omnipresent Excel spreadsheet that is featured on every computer: We use them even when it makes no sense— for example, when generating ten-year financial projections for start-ups or when comparing potential lovers that we have "sourced" from dating sites. Excel spreadsheets might as well be one of the most dangerous recent inventions.

Even in his own jurisdiction, the man with the hammer tends to overuse it. Literary reviewers are trained to detect authors' references, symbols, and hidden messages. As a novelist, I realize that literary reviewers conjure up such devices where there are none. This is not a million miles away from what business journalists do, too. They scour the most trivial utterings of central bank governors and somehow discover hints of fiscal policy change by parsing their words.

In conclusion: If you take your problem to an expert, don't expect the overall best solution. Expect an approach that can be solved with the expert's tool kit. The brain is not a central computer. Rather, it is a Swiss Army knife with many specialized tools. Unfortunately, our "pocketknives" are incomplete. Given our life experiences and our professional expertise, we already possess a few blades. But to better equip ourselves, we must try to add two or three additional tools to our repertoire—mental models that are far afield from our areas of expertise. For example, over the past few years, I have begun to take a biological view of the world and have won a new understanding of complex systems. Locate your shortcomings and find suitable knowledge and methodologies to balance them. It takes about a year to internalize the most important ideas of a new field, and it's worth it: Your pocketknife will be bigger and more versatile, and your thoughts sharper.

93

Mission Accomplished
Zeigarnik Effect

Berlin, 1927: A group of university students and professors visit a restaurant. The waiter takes order upon order, including special requests, but does not bother to write anything down. This is going to end badly, they think. But, after a short wait, all diners receive exactly what they ordered. After dinner, outside on the street, Russian psychology student Bluma Zeigarnik notices that she has left her scarf behind in the restaurant. She goes back in, finds the waiter with the incredible memory, and asks him if he has seen it. He stares at her blankly. He has no idea who she is or where she sat. "How can you have forgotten?" she asks indignantly. "Especially with your super memory!" The waiter replies curtly: "I keep every order in my head—until it is served."

Zeigarnik and her mentor, Kurt Lewin, studied this strange behavior and found that all people function more or less like the waiter. We seldom forget uncompleted tasks; they persist in our consciousness and do not let up, tugging at us like little children, until we give them our attention. On the other hand, once we've completed a task and checked it off our mental list, it is erased from memory.

The researcher has lent her name to this: Scientists now speak of the *Zeigarnik effect*. However, in her investigation, she uncovered a few untidy outliers: Some people kept a completely clear head even if they had dozens of projects on the go. Only in recent years could Roy Baumeister and his research team at Florida State University shed light on this. He took students who were a few months away from their final examinations and split them into three groups. Group 1 had to focus on a party during the current semester. Group 2 had to concentrate on the exam. Group 3 had to focus on the exam and also create a detailed study plan. Then Baumeister asked students to complete words under time pressure. Some students saw "pa . . ." and filled in "panic," while others thought of "party" or "Paris." This was a clever method of finding out what was on each of their minds. As expected, group 1 had relaxed about the upcoming exam, while students in group 2 could think of nothing else. Most astonishing was the result from group 3. Although these students also had to focus on the upcoming exam, their minds were clear and free from anxiety. Further experiments confirmed this. Outstanding tasks gnaw at us only until we have a clear idea of how we will deal with them. Zeigarnik mistakenly believed that it was necessary to complete tasks to erase them from memory. But it's not; a good plan of action suffices.

David Allen, the author of a best-selling book aptly entitled *Getting Things Done*, argues that he has one goal: to have a head as clear as water. For this, you don't need to have your whole life sorted into tidy compartments. But it does mean that you need a detailed plan for dealing with the messier areas. This plan must be divided into step-by-step tasks and preferably written down. Only when this is done can your mind rest. The adjective "detailed" is important. "Organize my wife's birthday party" or "find a new job" are worthless. Allen forces his clients to split such projects into twenty to fifty individual tasks.

It's worth noting that Allen's recommendation seems to fly in the face of the *planning fallacy* (chapter 91): the more detailed our planning, the more we tend to overlook factors from the periphery that will derail our projects. But here is the rub: If you want peace of mind, go for Allen's approach. If you want the most accurate estimate on cost, benefit, and duration of a project, forgot your detailed plan and look up similar projects. If you want both, do both.

Fortunately, you can do all this yourself with the aid of a decidedly low-tech device. Place a notepad by your bed. The next time you cannot get to sleep, jot down outstanding tasks and how you will tackle them. This will silence the cacophony of inner voices. "You want to find God, but you're out of cat food, so create a plan to deal with it," says Allen. His advice is sound, even if you have already found God or have no cat.

94

The Boat Matters More Than the Rowing
Illusion of Skill

Why are there so few serial entrepreneurs—businesspeople who start successful companies one after the other? Of course, there's Steve Jobs and Richard Branson, but they represent a tiny minority. Serial entrepreneurs account for less than 1 percent of everyone who starts a company. Do they all retire to their private yachts after the first success just like Microsoft co-founder Paul Allen did? Surely not. True businesspeople possess too much get-up-and-go to lie on a beach chair for hours on end. Is it because they can't let go and want to cosset their firms until they turn sixty-five? No. Most founders sell their shares within ten years. Actually, you would assume that such self-starters who are blessed with talent, a good personal network, and a solid reputation would be well equipped to found numerous other start-ups. So why do they stop? They didn't stop. They just failed at succeeding. Only one answer makes sense: Luck plays a bigger role than skill does. No businessperson likes to hear this. When I first heard about the *illusion of skill*, my reaction was: "What, my success was a fluke?" At first, it sounds a little offensive, especially if you worked hard to get there.

Let's take a sober look at business success: How much of it comes down to luck, and how much is the fruit of hard work and distinct talent? The question is easily misunderstood. Of course, little is achieved without talent, and nothing is achieved without hard work. Unfortunately, neither skills nor toil and trouble are the key criteria for success. They are *necessary*—but not sufficient. How do we know this? There is a very simple test: When a person is successful for a long time—more than that, when they enjoy more success in the long run compared to less qualified people—then and only then is talent the essential element. This is not the case with company founders; otherwise, the majority of successful entrepreneurs would, after the first achievement, continue to found and grow second, third, and fourth start-ups.

What about corporate leaders? How important are they to the success of a company? Researchers have determined a set of traits deemed to be associated with "a strong CEO"— management procedures, strategic brilliance in the past, and so on. Then they measured the relationship between these behaviors, on the one hand, and the increase of the companies' values during the reign of these CEOs, on the other hand. The result: If you compare two companies at random, in 60 percent of cases, the stronger CEO leads the stronger company. In 40 percent of the cases, the weaker CEO leads the stronger company. This is only 10 percentage points more than no relationship at all. Kahneman said: "It's hard to imagine that people enthusiastically buy books written by business leaders who are, on average, only slightly better than the norm." Even Warren

Buffett thinks nothing of CEO deification: "A good managerial record . . . is far more a function of what business boat you get into than it is of how effectively you row."

In certain areas, skill plays no role whatsoever. In his book *Thinking, Fast and Slow*, Kahneman describes his visit to an asset management company. To brief him, they sent him a spreadsheet showing the performance of each investment adviser over the past eight years. From this, a ranking was assigned to each: number 1, 2, 3, and so on in descending order. This was compiled every year. Kahneman quickly calculated the relationship between the years' rankings. Specifically, he calculated the correlation of the rankings between year 1 and year 2, between year 1 and year 3, year 1 and year 4, up until year 7 and year 8. The result: pure coincidence. Sometimes the adviser was at the very top and sometimes the very bottom. If an adviser had a great year, this was neither bolstered by previous years nor carried into subsequent years. The correlation was zero. And yet the consultants pocketed bonuses for their performance. In other words, the company was rewarding luck rather than skill.

In conclusion: Certain people make a living from their abilities, such as pilots, plumbers, and lawyers. In other areas, skill is necessary but not critical, as with entrepreneurs and leaders. Finally, chance is the deciding factor in a number of fields, such as in financial markets. Here, the *illusion of skill* pervades. So, give plumbers due respect and chuckle at successful financial jesters.

95

Why Checklists Deceive You

Feature-Positive Effect

wo series of numbers: The first, series A, consists of: 724, 947, 421, 843, 394, 411, 054, 646. What do these numbers have in common? Don't read on until you have an answer. It's simpler than you think: The number 4 features in each of them. Now examine series B: 349, 851, 274, 905, 772, 032, 854, 113. What links these numbers? Do not read further until you've figured it out. Series B is more difficult, right? Answer: None use the number 6. What can you learn from this? Absence is much harder to detect than presence. In other words, we place greater emphasis on what is present than on what is absent.

Last week, while on a walk, it occurred to me that nothing hurt. It was an unexpected thought. I rarely experience pain anyway, but when I do, it is very present. But the absence of pain I rarely recognize. It was such a simple, obvious fact, it amazed me. For a moment, I was elated—until this little revelation slipped from my mind again.

At a classical recital, an orchestra performed Beethoven's Ninth Symphony. A storm of enthusiasm gripped the concert hall. During the ode in the fourth movement, tears of joy could be seen here and there. How fortunate we are that this symphony exists, I thought. But is that really true? Would we be less happy without the work? Probably not. Had the symphony never been composed, no one would miss it. The director would receive no angry calls saying: "Please have this symphony written and performed immediately." In short, what exists means a lot more than what is missing. Science calls this the *feature-positive effect*.

Prevention campaigns utilize this well. "Smoking causes lung cancer" is much more powerful than "Not smoking leads to a life free of lung cancer." Auditors and other professionals who employ checklists are prone to the *feature-positive effect*: Outstanding tax declarations are immediately obvious because they feature on their lists. What does not appear, however, is more artistic fraud, such as the goings-on at Enron and with Bernie Madoff's Ponzi scheme. Also absent are the undertakings of "rogue traders," such as Nick Leeson and Jerome Kerviel, to whom Barings and Société Générale fell victim. Financial vagaries of this kind are not on any checklist. And they do not have to be illegal: A mortgage bank will be on the lookout for credit risk due to a drop in the debtor's income because this appears on its list; however, it will overlook the devaluation of property, say, through the construction of an incineration plant in the vicinity.

Suppose you manufacture a dubious product, such as a salad dressing with a high level of cholesterol. What do you do? On the label, you promote the twenty different vitamins in the dressing and omit the cholesterol level. Consumers won't notice its absence. And the positive, present features will make sure that they feel safe and informed.

In academia, we constantly encounter the *feature-positive effect*. The confirmation of hypotheses leads to publications, and in exceptional cases these are rewarded with Nobel Prizes. On the other hand, the falsification of a hypothesis is a lot harder to get published, and as far as I know, there has never been a Nobel Prize awarded for this. However, such falsification is as scientifically valuable as confirmation. Another consequence of the effect is that we are also much more open to positive advice (do X) than to negative suggestions (forget about Y)—no matter how useful the latter may be.

In conclusion: We have problems perceiving nonevents. We are blind to what does not exist. We realize if there is a war, but we do not appreciate the absence of war during peacetime. If we are healthy, we rarely think about being sick. Or, if we get off the plane in Cancún, we do not stop to notice that we did not crash. If we thought more frequently about absence, we might well be happier. But it is tough mental work. The greatest philosophical question is: Why does something and not *nothing* exist? Don't expect a quick answer; rather, the question itself represents a useful instrument for combating the *feature-positive effect*.

96

Drawing the Bull's-Eye around the Arrow
Cherry Picking

On their websites, hotels present themselves in the very best light. They carefully select each photo, and only beautiful, majestic images make the cut. Unflattering angles, dripping pipes, and drab breakfast rooms are swept under the tattered carpet. Of course, you know this is true. When you are confronted by the shabby lobby for the first time, you simply shrug your shoulders and head to the registration desk.

What the hotel did is called *cherry picking*: selecting and showcasing the most attractive features and hiding the rest. As with the hotel experience, you approach other things with the same muted expectations: brochures for cars, real estate, or law firms. You know how they work, and you don't fall for them.

However, you respond differently to the annual reports of companies, foundations, and government organizations. Here, you tend to expect objective depictions. You are mistaken. These bodies also *cherry-pick*: If goals are achieved, they are talked up; if they falter, they are not even mentioned.

Suppose you are the head of a department. The board invites you to present your team's state of play. How do you tackle this? You devote most of your PowerPoint slides to elaborate on the team's triumphs and throw in a token few to identify "challenges." Any other unmet achievements you conveniently forget.

Anecdotes are a particularly tricky sort of *cherry picking*. Imagine you are the managing director of a company that manufactures some kind of technical device. A survey has revealed that the vast majority of customers cannot operate your gadget. It's too complicated. Now the HR manager gives his two cents, proclaiming: "My father-in-law picked it up yesterday and figured out how to work it right away." How much weight would you attach to this particular cherry? Right: close to zero. To rebuff an anecdote is difficult because it is a mini-story, and we know how vulnerable our brains are to those. To prevent this, cunning leaders train themselves throughout their careers to be hypersensitive to such anecdotes and to shoot them down as soon as they are uttered.

The more elevated or elite a field is, the more we fall for *cherry picking*. In *Antifragile*, Taleb describes how all areas of research—from philosophy to medicine to economics—brag about their results: "Like politicians, academia is well equipped to tell us what it did for us, not what it did not—hence it shows how indispensable her methods are." Pure *cherry picking*. But our respect for academics is far too great for us to notice this.

Or consider the medical profession: To tell people that they should not smoke is the greatest medical contribution of the past sixty years—superior to all the research and medical advances since the end of the Second World War. Physician Druin Burch confirms this in his book *Taking the Medicine*. A few cherries—antibiotics, for instance—distract us, and so drug researchers are celebrated while antismoking activists are not.

Administrative departments in large companies glorify themselves like hoteliers do. They are masters at showcasing all they have done, but they never communicate what they haven't achieved for the company. What should you do? If you sit on the supervisory board of such an organization, ask about the "leftover cherries," the failed projects and missed goals. You learn a lot more from this than from the successes. It is amazing how seldom such questions are asked. Second: Instead of employing a horde of financial controllers to calculate costs to the nearest cent, double-check targets. You will be amazed to find that, over time, the original goals have faded. These have been replaced, quietly and secretly, with self-set goals that are always attainable. If you hear of such targets, alarm bells should sound. It is the equivalent of shooting an arrow and drawing a bull's-eye around where it lands.

97

The Stone Age Hunt for Scapegoats
Fallacy of the Single Cause

Chris Matthews is one of MSNBC's top journalists. In his news show, so-called political experts are wheeled in one after the other and interviewed. I've never understood what a political expert is or why such a career is worthwhile. In 2003, the U.S. invasion of Iraq was the issue on everybody's lips. More important than the experts' answers were Chris Matthews's questions: "What is *the motive* behind the war?" "I wanted to know whether 9/11 is *the reason*, because a lot of people think it's payback." "Do you think that the weapons of mass destruction was *the reason* for this war?" "Why do you think we invaded Iraq? The *real reason*, not the sales pitch." And so on.

I can't abide questions like that anymore. They are symptomatic of the most common of all mental errors, a mistake for which, strangely enough, there is no everyday term. For now, the awkward phrase, the *fallacy of the single cause*, will have to do.

Five years later, in 2008, panic reigned in the financial markets. Banks caved in and had to be nursed back to health with tax dollars. Investors, politicians, and journalists probed furiously for the root of the crisis: Greenspan's loose monetary policy? The stupidity of investors? The dubious rating agencies? Corrupt auditors? Bad risk models? Pure greed? Not a single one, and yet every one of these, is the cause.

A balmy Indian summer, a friend's divorce, the First World War, cancer, a school shooting, the worldwide success of a company, the invention of writing—any clear-thinking person knows that no single factor leads to such events. Rather, there are hundreds, thousands, an infinite number of factors that add up. Still, we keep trying to pin the blame on just one.

"When an apple ripens and falls—what makes it fall? Is it that it is attracted to the ground, is it that the stem withers, is it that the sun has dried it up, that is has grown heavier, that the wind shakes it, that the boy standing underneath it wants to eat it? No one thing is the cause." In this passage from *War and Peace*, Tolstoy hit the nail on the head.

Suppose you are the product manager for a well-known breakfast cereal brand. You have just launched an organic, low-sugar variety. After a month, it's painfully clear that the new product is a flop. How do you go about investigating the cause? First, you know that there will never be one sole factor. Take a sheet of paper and sketch out all the potential reasons. Do the same for the reasons behind these reasons. After a while, you will have a network of possible influencing factors. Second, highlight those you can change and delete those you cannot (such as "human nature"). Third, conduct empirical tests by varying the highlighted factors in different markets. This costs time and money, but it's the only way to escape the swamp of superficial assumptions.

The *fallacy of the single cause* is as ancient as it is dangerous. We have learned to see people as the "masters of their own destinies." Aristotle proclaimed this 2,500 years ago. Today we know that it is wrong. The notion of free will is up for debate. Our actions are brought about by the interaction of thousands of factors—from genetic predisposition to upbringing, from education to the concentration of hormones between individual brain cells. Still we hold firmly to the old image of self-governance. This is not only wrong but also morally questionable. As long as we believe in singular reasons, we will always be able to trace triumphs or disasters back to individuals and stamp them "responsible." The idiotic hunt for a scapegoat goes hand in hand with the exercise of power—a game that people have been playing for thousands of years.

And yet the *fallacy of the single cause* is so popular that Tracy Chapman was able to build her worldwide success on it. "Give Me One Reason" is the song that secured her success. But hold on—weren't there a few others, too?

98

Why Speed Demons Appear to Be Safer Drivers
Intention-to-Treat Error

You'll find it hard to believe, but speed demons drive more safely than so-called careful drivers. Why? Well, consider this: The distance from Miami to West Palm Beach is around seventy-five miles. Drivers who cover the distance in an hour or less we'll categorize as "reckless drivers" because they're traveling at an average of 75 mph or more. All others we put into the group of careful drivers. Which group experiences fewer accidents? Without a doubt, it is the "reckless drivers." They all completed the journey in less than an hour, so they could not have been involved in any accidents. This automatically puts all drivers who end up in accidents in the slower drivers' category. This example illustrates a treacherous fallacy, the so-called *intention-to-treat error.* Unfortunately, there is no catchier term for it.

This might sound to you like the *survivorship bias* (chapter 1), but it's different. In the *survivorship bias* you see only the survivors, not the failed projects or cars involved in accidents. In the *intention-to-treat error*, the failed projects or cars with accidents prominently show up, just in the wrong category.

A banker showed me an interesting study recently. Its conclusion: Companies with debt on their balance sheets are significantly more profitable than firms with no debt (equity only). The banker vehemently insisted that every company should borrow at will, and, of course, his bank is the best place to do it. I examined the study more closely. How could that be? Indeed, from one thousand randomly selected firms, those with large loans displayed higher returns not only on their equity but also on their total capital. They were in every respect more successful than the independently financed firms. Then the penny dropped: Unprofitable companies don't get corporate loans. Thus, they form part of the "equity-only" group. The other firms that make up this set have bigger cash cushions, stay afloat longer, and, no matter how sickly they are, remain part of the study. On the other side, firms that have borrowed a lot go bankrupt more quickly. Once they cannot pay back the interest, the bank takes over, and the companies are sold off—thus disappearing from the sample. The ones that remain in the "debt group" are relatively healthy, regardless of how much debt they have amassed on their balance sheets.

If you're thinking, "Okay, got it," watch out. The *intention-to-treat error* is not easy to recognize. A fictional example from medicine: A pharmaceutical company has developed a new drug to fight heart disease. A study "proves" that it significantly reduces patients' mortality rates. The data speaks for itself: Among patients who have taken the drug regularly, the five-year mortality rate is 15 percent. For those who have swallowed placebo pills, it is about the same, indicating that the pill doesn't work. However—and this is crucial—the mortality rate of patients who have taken the drug at irregular intervals is 30 percent—twice as high! A big difference between regular and irregular intake. So, the pill is a complete success. Or is it?

Here's the snag: The pill is probably not the decisive factor; rather, it is the patients' behavior. Perhaps patients discontinued the pill following severe side effects and thus landed in the "irregular intake" category. Maybe they were so ill that there was no way to continue it on a regular basis. Either way, only relatively healthy patients remain in the "regular" group, which makes the drug look a lot more effective than it really is. The really sick patients who, for this very reason, couldn't take the drug on a regular basis ended up populating the "irregular intake" group.

In reputable studies, medical researchers evaluate the data of all patients whom they originally intend to treat (hence the title); it doesn't matter if they take part in the trial or they drop out. Unfortunately, many studies flout this rule. Whether this is intentional or accidental remains to be seen. Therefore, be on your guard: Always check whether test subjects—drivers who end up in accidents, bankrupt companies, critically ill patients—have, for whatever reason, vanished from the sample. If so, you should file the study where it belongs: in the trash can.

99

Why You Shouldn't Read the News

News Illusion

Earthquake in Sumatra. Plane crash in Russia. Man holds daughter captive in cellar for thirty years. Heidi Klum separates from Seal. Record salaries at Bank of America. Attack in Pakistan. Resignation of Mali's president. New world record in shot put.

Do you really need to know all these things?

We are incredibly well informed, yet we know incredibly little. Why? Because two centuries ago, we invented a toxic form of knowledge called "news." News is to the mind what sugar is to the body: appetizing, easy to digest—and highly destructive in the long run.

Three years ago, I began an experiment. I stopped reading and listening to the news. I canceled all newspaper and magazine subscriptions. Television and radio were disposed of. I deleted the news apps from my iPhone. I didn't touch a single free newspaper and deliberately looked the other way when someone on a plane tried to offer me any such reading material. The first weeks were hard. Very hard. I was constantly afraid of missing something. But after a while, I had a new outlook. The result after three years: clearer thoughts, more valuable insights, better decisions, and much more time. And the best thing? I haven't missed anything important. My social network—not Facebook, the one that exists in the real world consisting of flesh-and-blood friends and acquaintances—works as a news filter and keeps me in the loop.

A dozen reasons exist to give news a wide berth. Here are the top three: First, our brains react disproportionately to different types of information. Scandalous, shocking, people-based, loud, fast-changing details all stimulate us, whereas abstract, complex, and unprocessed information sedates us. News producers capitalize on this. Gripping stories, garish images, and sensational "facts" capture our attention. Recall for a moment their business models: Advertisers buy space and thus finance the news circus on the condition that their ads will be seen. The result: Everything subtle, complex, abstract, and profound must be systematically filtered out, even though such stories are much more relevant to our lives and to our understanding of the world. As a result of news consumption, we walk around with a distorted mental map of the risks and threats we actually face.

Second, news is irrelevant. In the past twelve months, you have probably consumed about ten thousand news snippets—perhaps as many as thirty per day. Be very honest: Name one of them, just one that helped you make a better decision—for your life, your career, or your business—compared with not having this piece of news. No one I have asked has been able to name more than two useful news stories—out of ten thousand. A miserable result. News organizations assert that their information gives you a competitive advantage. Too many fall for this. In reality, news consumption represents a competitive disadvantage. If news really helped people advance, journalists would be at the top of the income pyramid. They aren't—quite the opposite.

Third, news is a waste of time. An average human being squanders half a day each week on reading about current affairs. In global terms, this is an immense loss of productivity. Take the 2008 terror attacks in Mumbai. Out of sheer thirst for recognition, terrorists murdered two hundred people. Let's say a billion people devoted an hour of their time to following the aftermath: They viewed the minute-by-minute updates and listened to the inane chatter of a few "experts" and "commentators." This is a very realistic "guesstimate" since India has more than a billion inhabitants. Thus our conservative calculation: One billion people multiplied by an hour's distraction equals one billion hours of work stoppage. If we convert this, we learn that news consumption wasted around two thousand lives—ten times more than the attack. A sarcastic but accurate observation.

I would predict that turning your back on news will benefit you as much as purging any of the other ninety-eight flaws we have covered in the pages of this book. Kick the habit—completely. Instead, read long background articles and books. Yes, nothing beats books for understanding the world.

EPILOGUE

The pope asked Michelangelo: "Tell me the secret of your genius. How have you created the statue of David, the masterpiece of all masterpieces?" Michelangelo's answer: "It's simple. I removed everything that is not David."

Let's be honest. We don't know for sure what makes us successful. We can't pinpoint exactly what makes us happy. But we know with certainty what destroys success or happiness. This realization, as simple as it is, is fundamental: Negative knowledge (what *not* to do) is much more potent than positive knowledge (what to do).

Thinking more clearly and acting more shrewdly means adopting Michelangelo's method: Don't focus on David. Instead, focus on everything that is not David and chisel it away. In our case: Eliminate all errors and better thinking will follow.

The Greeks, Romans, and medieval thinkers had a term for this approach: *via negativa*. Literally, the negative path, the path of renunciation, of exclusion, of reduction. Theologians were the first to tread the *via negativa*: We cannot say what God is; we can only say what God is not. Applied to the present day: We cannot say what brings us success. We can pin down only what blocks or obliterates success. Eliminate the downside, the thinking errors, and the upside will take care of itself. This is all we need to know.

As a novelist and company founder, I have fallen into a variety of traps. Fortunately I was always able to free myself from them. Nowadays when I hold presentations in front of doctors, CEOs, board members, investors, politicians, or government officials, I sense a kinship. I feel that we are sitting in the same boat—after all, we are all trying to row through life without getting swallowed up by the maelstroms. Still, many people are uneasy with the *via negativa*. It is counterintuitive. It is even countercultural, flying in the face of contemporary wisdom. But look around and you'll find plenty of examples of the *via negativa* at work. This is what the legendary investor Warren Buffett writes about himself and his partner Charlie Munger: "Charlie and I have not learned how to solve difficult business problems. What we have learned is to avoid them." Welcome to the *via negativa*.

I have listed almost one hundred thinking errors in this book without answering the question: What are thinking errors anyway? What is irrationality? Why do we fall into these traps? Two theories of irrationality exist: a *hot* and a *cold*. The *hot* theory is as old as the hills. Here is Plato's analogy: A rider steers wildly galloping horses; the rider signifies reason and the galloping horses embody emotions. Reason tames feelings. If this fails, irrationality runs free. Another example: Feelings are like bubbling lava. Usually, reason can keep a lid on them, but every now and then the lava of irrationality erupts. Hence *hot* irrationality. There is no reason to fret about logic: It is error-free; it's just that, sometimes, emotions overpower it.

This hot theory of irrationality boiled and bubbled for centuries. For John Calvin, the founder of a strict form of Protestantism in the 1500s, such feelings represented evil, and only by focusing on God could you repel them. People who underwent volcanic eruptions of emotion were of the devil. They were tortured and killed. According to Austrian psychoanalyst Sigmund Freud's theory, the rationalist "ego" and the moralistic "superego" control the impulsive "id." But that theory holds less water in the real world. Forget about obligation and discipline. To believe that we can completely control our emotions through thinking is illusory—as illusory as trying to make your hair grow by willing it to.

On the other hand, the *cold* theory of irrationality is still young. After the Second World War, many searched for explanations about the irrationality of the Nazis. Emotional outbursts were rare in Hitler's leadership ranks. Even his fiery speeches were nothing more than masterful performances. It was not molten eruptions but stone-cold calculation that resulted in the Nazi madness. The same can be said of Stalin or of the Khmer Rouge.

In the 1960s, psychologists began to do away with Freud's claims and to examine our thinking, decisions, and actions scientifically. The result was a cold theory of irrationality that states: Thinking is in itself not pure, but prone to error. This affects everyone. Even highly intelligent people fall into the same cognitive traps. Likewise, errors are not randomly distributed. We systematically err in the same direction. That makes our mistakes predictable, and thus fixable to a degree—but only to a degree, never completely. For a few decades, the origins of these errors remained in the dark. Everything else in our body is relatively reliable—heart, muscles, lungs, immune system. Why should our brains of all things experience lapse after lapse?

Thinking is a biological phenomenon. Evolution has shaped it just as it has the forms of animals or the colors of flowers. Suppose we could go back fifty thousand years, grab hold of an ancestor, and bring him back with us into the present. We send him to the hairdresser and put him in a Hugo Boss suit. Would he stand out on the street? No. Of course, he would have to learn English, how to drive, and how to operate a cell phone, but we had to learn those things, too. Biology has dispelled all doubt: Physically, and that includes cognitively, we are hunter-gatherers in Hugo Boss (or H&M, as the case may be).

What has changed markedly since ancient times is the environment in which we live. Back then, things were simple and stable. We lived in small groups of about fifty people. There was no significant technological or social progress. Only in the last ten thousand years did the world begin to transform dramatically, with the development of crops, livestock, villages, cities, global trade, and financial markets. Since industrialization, little is left of the environment for which our brain is optimized. If you spend fifteen minutes in a shopping mall, you will pass more people than our ancestors saw during their entire lifetimes. Whoever claims to know how the world will look in ten years is made into a laughingstock less than a year after such a pronouncement. In the past ten thousand years, we have created a world that we no longer understand. Everything is more sophisticated, but also more complex and interdependent. The result is overwhelming material prosperity, but also lifestyle diseases (such as type 2 diabetes, lung cancer, and depression) and errors in thinking. If the complexity continues to rise—and it will, that much is certain—these errors will only increase and intensify.

In our hunter-gatherer past, activity paid off more often than reflection did. Lightning-fast reactions were vital, and long ruminations were ruinous. If your hunter-gatherer buddies suddenly bolted, it made sense to follow suit—regardless of whether a saber-toothed tiger or a boar had startled them. If you failed to run away, and it turned out to be a tiger, the price of a first-degree error was death. On the other hand, if you had just fled from a board, this lesser mistake would have cost you only a few calories. It paid to be wrong about the same things. Whoever was wired differently exited the gene pool after the first or second incidence. We are the descendants of those *homines sapientes* who tend to flee when the crowd does. But in the modern world, this intuitive behavior is disadvantageous. Today's world rewards single-minded contemplation and independent action. Anyone who has fallen victim to stock market hype has witnessed that.

Evolutionary psychology is still mostly a theory, but a very convincing one at that. It explains the majority of flaws, though not all of them. Consider the following statement: "Every Hershey bar comes in a brown wrapper. Thus, every candy bar in a brown wrapper must be a Hershey bar." Even intelligent people are susceptible to this flawed conclusion—so are native tribes that, for the most part, remain untouched by civilization. Our hunter-gatherer ancestors were certainly not impervious to faulty logic. Some bugs in our thinking are hardwired and have nothing to do with the "mutation" of our environment.

Why is that? Evolution does not "optimize" us completely. As long as we advance beyond our competitors (i.e., beat the

Neanderthals), we can get away with error-laced behavior. Consider the cuckoo: For hundreds of thousands of years, they have laid their eggs in the nests of songbirds, which then incubate and even feed the cuckoo chicks. This represents a behavioral error that evolution has not erased from the smaller birds; it is not deemed to be serious enough.

A second, parallel explanation of why our mistakes are so persistent took shape in the late 1990s: Our brains are designed to reproduce rather than search for the truth. In other words, we use our thoughts primarily to persuade. Whoever convinces others secures power and thus access to resources. Such assets represent a major advantage for mating and for rearing offspring. That truth is, at best, a secondary focus and is reflected in the book market: Novels sell much better than nonfiction titles, in spite of the latter's superior candor.

Finally, a third explanation exists: Intuitive decisions, even if they lack logic, are better under certain circumstances. So-called heuristic research deals with this topic. For many decisions, we lack the necessary information, so we are forced to use mental shortcuts and rules of thumb (heuristics). If you are drawn to different potential romantic partners, you must evaluate whom to marry. This is not a rational decision; if you rely solely on logic, you will remain single forever. In short, we often decide intuitively and justify our choices later. Many decisions (career, life partner, investments) take place subconsciously. A fraction of a second later, we construct a reason so that we feel we made a conscious choice. Alas, we do not behave like scientists who are purely interested in objective facts. Instead, we think like lawyers, crafting the best possible justification for a predetermined conclusion.

So, forget about the "left and right brain" that semi-intelligent self-help books describe. Much more important is the difference between intuitive and rational thinking. Both have legitimate applications. The intuitive mind is swift, spontaneous, and energy-saving. Rational thinking is slow, demanding, and energy-guzzling (in the form of blood sugar). Nobody has described this better than the great Daniel Kahneman in *Thinking, Fast and Slow*.

Since I started to collect cognitive errors, people often ask me how I manage to live an error-free life. The answer is: I don't. In fact, I don't even try. Just like everybody else, I make snap decisions by consulting not my thoughts but my feelings. For the most part I substitute the question "What do I think about this?" with "How do I feel about this?" Quite frankly, anticipating and avoiding fallacies is a costly undertaking.

To make things simple, I have set myself the following rules: In situations where the possible consequences are large (i.e., important personal or business decisions), I try to be as reasonable and rational as possible when choosing. I take out my list of errors and check them off one by one, just like a pilot does. I've created a handy checklist decision tree, and I use it to examine important decisions with a fine-tooth comb. In situations where the consequences are small (i.e., regular or Diet Pepsi, sparkling or flat water?), I forget about rational optimization and let my intuition take over. Thinking is tiring. Therefore, if the potential harm is small, don't rack your brains; such errors won't do lasting damage. You'll live better like this. Nature doesn't seem to mind if our decisions are perfect or not, as long as we can maneuver ourselves through life—and as long as we are ready to be rational when it comes to the crunch. And there's one other area where I let my intuition take the lead: when I am in my "circle of competence." If you practice an instrument, you learn the notes and tell your fingers how to play them. Over time, you know the keys or the strings inside out. You see a musical score and your hands play the notes almost automatically. Warren Buffett reads balance sheets like professional musicians read scores. This is his circle of competence, the field he intuitively understands and masters. So, find out where your circle of competence is. Get a clear grasp of it. Hint: It's smaller than you think. If you face a consequential decision outside that circle, apply the hard, slow, rational thinking. For everything else, give your intuition free rein.

ACKNOWLEDGMENTS

Thanks to my friend Nassim Nicholas Taleb for inspiring me to write this book, even if his advice was not to publish it under any circumstances. Alas, he encouraged me to write novels, arguing that nonfiction isn't "sexy." The time we have passed together discussing how to live in a world we don't understand have been my favorite hours of the week. Thanks to Koni Gebistorf, who masterfully edited the original German texts, and to Nicky Griffin, who translated the book into English (when she was away from her office at Google). I couldn't have picked better publishers and editors than Hollis Heimbouch from HarperCollins and Drummond Moir from Sceptre who have given these chapters their final finesse. Thanks to the scientists of the ZURICH.MINDS community for the countless debates about the state of research. Special thanks go to Gerd Gigerenzer, Roy Baumeister, Leda Cosmides, John Tooby, Robert Cialdini, Jonathan Haidt, Ernst Fehr, Bruno Frey, Iris Bohnet, Dan Golstein, Tomáš Sedláček and the philosopher John Gray for the enlightening conversations. I also thank my literary agent, John Brockman, and his superb crew, for helping me with both the American and British editions of this book. Thanks to Frank Schirrmacher for

Acknowledgments

finding space for my columns in the *Frankfurter Allgemeine Zeitung*, to Giovanni di Lorenzo and Moritz Mueller-Wirth for their publication in *Die Zeit* (Germany), and to Martin Spieler who gave them a good home in Switzerland's *Sonntagszeitung*. Without the weekly pressure to forge one's thoughts into a readable format, my notes would never have been published in book form.

For everything that appears here after the countless stages of editing, I alone bear the responsibility. My greatest thanks goes to my wife, Sabine Ried, who proves to me every day that the "good life"—as defined by Aristotle—consists of far more than clear thoughts and clever actions.

A NOTE ON SOURCES

Hundreds of studies have been conducted on the vast majority of cognitive and behavioral errors. In a scholarly work, the complete reference section would easily double the pages of this book. I have focused on the most important quotes, technical references, recommendations for further reading, and comments. The knowledge encompassed in this book is based on the research carried out in the fields of cognitive and social psychology over the past three decades.

SURVIVORSHIP BIAS
Survivorship bias in funds and stock market indices, see: Edwin J. Elton, Martin J. Gruber, and Christopher R. Blake, "Survivorship Bias and Mutual Fund Performance," *The Review of Financial Studies* 9, no. 4 (1996): 1097–1120.
Statistically relevant results by coincidence (self-selection), see: John P. A. Ioannidis, "Why Most Published Research Findings Are False," *PLoS Med* 2, no. 8 (2005): e124.

SWIMMER'S BODY ILLUSION
Nassim Nicholas Taleb, *The Black Swan: The Impact of the Highly Improbable* (New York: Random House, 2007), 109–10.
"Ideally, the comparison should be made between people who went to Harvard and people who were admitted to Harvard but chose instead to go to Podunk State. Unfortunately, this is likely to produce samples too small for statistical analysis." Thomas Sowell, *Economic Facts and Fallacies* (New York: Basic Books, 2008), 106.
David Lykken and Auke Tellegen, "Happiness Is a Stochastic Phenomenon," *Psychological Science* 7, no. 3 (May 1996): 189.

In his book *Good to Great*, Jim Collins cites the CEO of Pitney Bowes, Dave Nassef: "I used to be in the Marines, and the Marines get a lot of credit for building people's values. But that's not the way it really works. The Marine Corps recruits people who share the corps' values, then provides them with training required to accomplish the organization's mission."

CLUSTERING ILLUSION

The random sequence OXXXOXXXXOXXOOOXOOXXOO: Thomas Gilovich, *How We Know What Isn't So: The Fallibility of Human Reason in Everyday Life* (New York: Free Press, 1993), 16.

Daniel Kahneman and Amos Tversky, "Subjective Probability: A Judgment of Representativeness," in Daniel Kahneman, Paul Slovic, and Amos Tversky, *Judgment under Uncertainty: Heuristics and Biases* (New York: Cambridge University Press, 1982), 32–47.

This paper caused controversy because it destroyed many athletes and sports commentators' belief in the "hot hand"—in lucky streaks: Thomas Gilovich, Robert Vallone, and Amos Tversky, "The Hot Hand in Basketball: On the Misperception of Random Sequences," *Cognitive Psychology* 17 (1985): 295–314.

The Virgin Mary on toast on BBC: accessed November 1, 2012, http://news.bbc.co.uk/2/hi/4034787.stm.

The *clustering illusion* has been recognized for centuries. In the eighteenth century, David Hume commented in *The Natural History of Religion*: "We see faces on the moon and armies in the clouds."

Other examples from the Wikipedia entry for "Perceptions of Religious Imagery in Natural Phenomena": The "Nun Bun" was a cinnamon pastry whose markings resembled the nose and jowls of Mother Teresa. It was found in a Nashville coffee shop in 1996 but was stolen on Christmas in 2005. See: "Mother Teresa Is Not Amused," *Seattle Times*, May 22, 1997. "Our Lady of the Underpass" was another appearance by the Virgin Mary, this time as a salt stain under Interstate 94 in Chicago in 2005. Other cases include Hot Chocolate Jesus, Jesus on a shrimp tail dinner, Jesus in a dental X-ray, and a Cheeto shaped like Jesus.

A side comment: I don't understand how people can recognize the face of Jesus—or of the Virgin Mary. Nobody knows how he looked. No pictures exist from his lifetime.

Recognizing faces in objects is called "pareidolia"—clocks, the front of a car, the moon.

The brain processes different things in different regions. As soon as an object looks like a face, the brain treats it like a face—this is very different from other objects.

SOCIAL PROOF

Robert B. Cialdini, *Influence: The Psychology of Persuasion*, rev. ed. (New York: William Morrow, 1993), 114–65.

Solomon E. Asch, "Effects of Group Pressure upon the Modification and Distortion of Judgment," in H. Guetzkow (ed.), *Groups, Leadership and Men* (Pittsburgh: Carnegie Press, 1951), 177–90.

Canned laughter works especially well if it's in-group laughter. "Participants laughed and smiled more, laughed longer, and rated humorous material more favorably when they heard in-group laughter rather than out-group laughter or no laughter at all." See: Michael J. Platow et al., "It's Not Funny If *They're* Laughing: Self-Categorization, Social Influence, and Responses to Canned Laughter," *Journal of Experimental Social Psychology* 41, no. 5 (2005): 542–50.

The storm of enthusiasm for Goebbels's speech did not stem from *social proof* alone. What you do not see in the YouTube video is a banner above the speaker declaring "Total War = Shortest War," an argument that made sense to many. After the Stalingrad debacle, people were sick of the war. Thus, the population had to be won back with this argument: The more aggressively it was fought, the quicker it would be over. Thanks to Johannes Grützig (Germany) for this insight. My comment: I don't think that before the speech the Hitler regime was interested in waging war for longer than was necessary. In this respect, Goebbels's argument is not convincing.

Besides the vacation restaurant, there's another case where social proof is of value: if you have tickets to a football game in a foreign city and don't know where the stadium is. Here, it makes sense to follow the people who look like football fans.

German philosopher Friedrich Nietzsche warned half a century before the Goebbel craze: "Madness is a rare thing in individuals—but in groups, parties, peoples, and ages it is the rule."

SUNK COST FALLACY

The classic research on the sunk cost fallacy is: H. R. Arkes and C. Blumer, "The Psychology of Sunk Cost," *Organizational Behavior and Human Decision Processes* 35 (1985): 124–40. In this research, Arkes and Blumer asked subjects to imagine that they had purchased tickets for a ski trip to Michigan (at a price of $100) and to Wisconsin (at a price of $50)—for the same day. The tickets are nonrefundable. Which ticket are you going to keep, assuming that you prefer the Wisconsin trip? Most subjects picked the less preferred trip to Michigan because of its higher ticket price.

A Note on Sources

On the Concorde, see: P. J. Weatherhead, "Do Savannah Sparrows Commit the Concorde Fallacy?," in *Behavioral Ecology and Sociobiology* (Berlin: Springer-Verlag, 1979), vol. 5, 373–81.

It's a strange finding that lower animals and children don't exhibit the *sunk cost fallacy*. Only in later years do we start to display this wrong behavior. Read: Hal R. Arkes and Peter Ayton, "The Sunk Cost and Concorde Effects: Are Humans Less Rational than Lower Animals?," *Psychological Bulletin* 125 (1999): 591–600.

RECIPROCITY

Robert B. Cialdini, *Influence: The Psychology of Persuasion*, rev. ed. (New York: HarperCollins, 1993), 17–56.

Robert Trivers published the theory of reciprocal altruism in 1971, which shed light on all kinds of human behavior. Thus, *reciprocity* is the basis for biological cooperation—besides kinship. See any basic biology textbook since 1980.

For evolutionary psychology's justification of reciprocity, see: David M. Buss, *Evolutionary Psychology: The New Science of the Mind* (Boston: Allyn and Bacon, 1999). Also: Roy F. Baumeister, *The Cultural Animal: Human Nature, Meaning, and Social Life* (Oxford, UK: Oxford University Press, 2005).

CONFIRMATION BIAS (PART 1)

How Darwin handled the *confirmation bias*, in: Charles T. Munger, *Poor Charlie's Almanack*, expanded 3rd ed. (Virginia Beach, VA: The Donning Company Publishers, 2006), 462.

"What Keynes was reporting is that the human mind works a lot like the human egg. When one sperm gets into a human egg, there's an automatic shut-off device that bars any other sperm from getting in. The human mind tends strongly toward the same sort of result. And so, people tend to accumulate large mental holdings of fixed conclusions and attitudes that are not often reexamined or changed, even though there is plenty of good evidence that they are wrong." In: Munger, *Poor Charlie's Almanack*, 461.

"What the human being is best of doing, is interpreting all new information so that their prior conclusions remain intact." Warren Buffett at the Berkshire Hathaway annual meeting, 2002, quoted in Peter Bevelin, *Seeking Wisdom: From Darwin to Munger* (Malmö, Sweden: PCA Publications, 2007), 56.

Nassim Nicholas Taleb, *The Black Swan: The Impact of the Highly Improbable* (New York: Random House, 2007), 58–59.

For the experiment with the sequence of numbers, see: Peter C. Wason, "On the Failure to Eliminate Hypotheses in a Conceptual Task," *Quarterly Journal of Experimental Psychology* 12, no. 3 (1960): 129–40.

"Faced with the choice between changing one's mind and proving there is no need to do so, almost everyone gets busy on the proof." John Kenneth Galbraith, *The Essential Galbraith* (New York: Houghton Mifflin, 2001), 241.

CONFIRMATION BIAS (PART 2)

Stereotyping as a special case of the *confirmation bias*, see: Roy F. Baumeister, *The Cultural Animal: Human Nature, Meaning, and Social Life* (Oxford, UK: Oxford University Press, 2005), 198–200.

AUTHORITY BIAS

Robert B. Cialdini, *Influence: The Psychology of Persuasion*, rev. ed. (New York: HarperCollins, 1993), 208–36.

For the track record of doctors before 1900 and a beautiful exposition on the authority of doctors and their strange theories, see: Noga Arkiha, *Passions and Tempers: A History of the Humours* (New York: Harper Perennial, 2008).

"Iatrogenic" conditions and injuries are those caused by medical treatment, for example, bloodletting.

After the 2008 financial crisis, two unexpected events of global proportions (*Black Swans*) took place: The Arab uprisings (2011) and the tsunami/ nuclear disaster in Japan (2011). Not one of the world's estimated 100,000 political and security authorities foresaw (or even could foresee) these events. This should be reason enough to distrust them—particularly if they are "experts" in all things social (fashion trends, politics, economics). These people are not stupid. They are simply misfortunate enough to have chosen a career in which they cannot win. Two alternatives are open to them: (a) to admit they don't know (not the best choice if you have a family to feed) or (b) to spout hot air.

Stanley Milgram, *Obedience to Authority; An Experimental View* (New York: Harper and Row, 1974). There is also a great DVD entitled *Obedience* (1969).

"If a CEO is enthused about a particularly foolish acquisition, both his internal staff and his outside advisors will come up with whatever projections are needed to justify his stance. Only in fairy tales are emperors told that they are naked." In: Warren Buffett, letter to shareholders of Berkshire Hathaway, 1998.

A Note on Sources

CONTRAST EFFECT
Robert B. Cialdini, *Influence: The Psychology of Persuasion*, rev. ed. (New York: HarperCollins, 1993), 11–16.

Charlie Munger calls the *contrast effect* the "Contrast-Misreaction Tendency." See: Charles T. Munger, *Poor Charlie's Almanack*, expanded 3rd ed. (Virginia Beach, VA: The Donning Company Publishers, 2006), 483.

Dan Ariely refers to the effect as the "relativity problem." See: Dan Ariely, *Predictably Irrational: The Hidden Forces That Shape Our Decisions*, rev. and expanded ed. (New York: Harper, 2009), chapter 1.

Contrasting factors may lead you to take the long way around: See: Daniel Kahneman and Amos Tversky, "Prospect Theory: An Analysis of Decision under Risk," *Econometrica* 47, no. 2 (1979): 263–92.

AVAILABILITY BIAS
The example with the letter "k": Amos Tversky and Daniel Kahneman, "Availability: A Heuristic for Judging Frequency and Probability," *Cognitive Psychology* 5 (1973): 207–32.

The *availability bias* leads to a wrong risk map in our mind. Tornadoes, airplane crashes, and electrocutions are widely reported in the media, which makes them easily available in our minds. On the other hand, deaths resulting from asthma, vaccinations, and glucose intolerance are underestimated because they are usually not reported. Read: Sarah Lichtenstein et al., "Judged Frequency of Lethal Events," *Journal of Experimental Psychology: Human Learning and Memory* 4 (1978): 551–78.

Another great quote from Charlie Munger on the *availability bias*: "You see that again and again—that people have some information they can count well and they have other information much harder to count. So they make the decision based only on what they can count well. And they ignore much more important information because its quality in terms of numeracy is less—even though it's very important in terms of reaching the right cognitive result. All I can tell you is that around Wesco [Charlie Munger's investment firm, comment RD] and Berkshire, we try not to be like that. We have Lord Keynes' attitude, which Warren quotes all the time: 'We'd rather be roughly right than precisely wrong.' In other words, if something is terribly important, we'll guess at it rather than just make our judgment based on what happens to be easily countable." In: Peter Bevelin, *Seeking Wisdom: From Darwin to Munger* (Malmö, Sweden: PCA Publications, 2007), 176.

Another way of stating the *availability bias* by Charlie Munger: "An idea or a fact is not worth more merely because it is easily available to you." In:

Charles T. Munger, *Poor Charlie's Almanack*, expanded 3rd ed. (Virginia Beach, VA: The Donning Company Publishers, 2006), 486. Quoted from Wesco Financial annual meeting, 1990, *Outstanding Investor Digest*, June 28, 1990, 20–21.

The *availability bias* is the reason why, when it comes to risk management, firms focus primarily on risks in the financial market: There is plenty of data on this. With operational risk, however, there is almost no data. It's not public. You would have to painstakingly cobble it together from many companies and that's expensive. For this reason, we create theories using material that is easy to find.

"The medical literature shows that physicians are often prisoners of their first-hand experience: their refusal to accept even conclusive studies is legendary." Robyn M. Dawes, *Everyday Irrationality: How Pseudo-Scientists, Lunatics, and the Rest of Us Systematically Fail to Think Rationally* (New York: Westview Press, 2011), 102.

Confidence in the quality of your own decisions depends solely on the number of decisions (predictions) made, regardless of how accurate or inaccurate they were. This is the chief problem with consultants. They make tons of decisions and predictions, but seldom validate them after the fact. They are on to the next projects, the next clients, and if something went wrong, well, it was a faulty implementation of their ideas and strategies. See: Hillel J. Einhorn and Robin M Hogarth, "Confidence in Judgment: Persistence of the Illusion of Validity," *Psychological Review* 85, no. 5 (September 1978): 395–416.

THE IT'LL-GET-WORSE-BEFORE-IT-GETS-BETTER FALLACY

No reference literature. This error in thinking is obvious.

STORY BIAS

"The king died and then the queen" is a story. "The king died and then the queen died of grief" is a plot. The difference between the two is causality. The English novelist E. M. Forster proposed this distinction in 1927.

Scientists still debate about which version of the king/queen debate is easier to recall from memory. The results of one study point to the following direction: If it takes a lot of mental effort to link two propositions, then recall is poor. If it takes zero mental effort to link two propositions, recall is poor, too. But if it takes an intermediate level of mental work, then recall is best. In other words, take these two sentences: "Joey's big brother punched him again and again. The next day his body was covered by bruises." "Joey's crazy mother became furiously angry with him. The next day his body was covered by bruises." To understand the second pair of sentences, you must make an extra logical inference. By putting in this extra work you form a richer memory for what you've read. The following study showed that recognition and recall memory for the causes was poorest for the most and least related causes and best for causes of intermediate levels of relatedness. Janice E. Keenan et al., "The Effects of Causal Cohesion on Comprehension and Memory," *Journal of Verbal Learning and Verbal Behavior* 23, no. 2 (April 1984): 115–26.

Robyn M. Dawes, *Everyday Irrationality: How Pseudo-Scientists, Lunatics, and the Rest of Us Systematically Fail to Think Rationally* (New York: Westview Press, 2001), 111–13.

"Narrative imagining—story—is the fundamental instrument of thought." Mark Turner, *The Literary Mind: The Origins of Thought and Language* (New York: Oxford University Press, 1998), 4.

The vignette of the car driving over the bridge, from Nassim Nicholas Taleb, personal communication.

HINDSIGHT BIAS

On Reagan's election: John F. Stacks, "Where the Polls Went Wrong," *Time* magazine, December 1, 1980.

One of the classic studies is from Baruch Fischhoff. He asked people to judge the outcome of a war they knew little about (British forces against the Nepalese Gurkhas in Bengal in 1814). Those who knew the outcome judged that outcome as much more probable. See: Baruch Fischhoff, "Hindsight ≠ Foresight: The Effect of Outcome Knowledge on Judgment under Uncertainty," *Journal of Experimental Psychology: Human Perception and Performance* 104 (1975): 288–99.

H. Blank, J. Musch, and R. Pohl, "Hindsight Bias: On Being Wise after the Event," *Social Cognition* 25, no. 1 (2007): 1–9.

OVERCONFIDENCE EFFECT

The original research paper on overconfidence: Sarah Lichtenstein and Baruch Fischhoff, "Do Those Who Know More Also Know More about How Much They Know?," *Organizational Behavior and Human Performance* 20 (1977): 159–83.

Marc Alpert and Howard Raiffa, "A Progress Report on the Training of Probability Assessors," in Daniel Kahneman, Paul Slovic, and Amos Tversky, *Judgment under Uncertainty: Heuristics and Biases* (New York: Cambridge University Press, 1982), 294–305.

Ulrich Hoffrage, "Overconfidence," in Rüdiger Pohl, *Cognitive Illusions: A Handbook on Fallacies and Biases in Thinking, Judgment and Memory* (Hove, UK: Psychology Press, 2004), 235–54.

Dale Griffin and Amos Tversky, "The Weighing of Evidence and the Determinants of Confidence," in Thomas Gilovich, Dale Griffin, and Daniel Kahneman (eds.), *Heuristics and Biases: The Psychology of Intuitive Judgment* (Cambridge, UK: Cambridge University Press, 2002), 230–49.

Even self-predictions are consistently overconfident: Robert P. Vallone, Dale W. Griffin, Sabrina Lin, and Lee Ross, "Overconfident Predictions of Future Actions and Outcomes by Self and Others," *Journal of Personality and Social Psychology* 58, no. 4 (April 1990): 582–92.

See also: Roy F. Baumeister, *The Cultural Animal: Human Nature, Meaning, and Social Life* (Oxford, UK: Oxford University Press, 2005), 241–44.

" . . . for men, overconfidence probably paid off more than underconfidence did." To learn more about why male *overconfidence* was important for evolution, see this interesting hypothesis: Roy F. Baumeister, *Is There Anything Good about Men? How Cultures Flourish by Exploiting Men* (Oxford, UK: Oxford University Press, 2010), 211–13.

Discussion on *overconfidence*, particularly the hypothesis that an inflated self-image benefits health, see: Scott Plous, *The Psychology of Judgment and Decision Making* (New York: McGraw-Hill, 1993), chapter 19, 217–30.

Extreme confidence or even overconfidence plays a role in the relationship between patient and doctor. "Doctors need to have some level of confidence to be able to interact with patients and everybody else, the nurses . . . In the emergency room, when everything is happening at once and the patient's in shock, I like to hear a voice that's steady and calm." Dr. Keating quoted in Christopher Chabris and Daniel Simons, *The Invisible Gorilla: And Other Ways Our Intuitions Deceive Us* (New York: Crown, 2010), 104.

"We all encounter hundreds or even thousands of people whom we don't know well, but whose confidence we can observe—and draw conclusions from. For such casual acquaintances, confidence is a weak signal. But in a smaller-scale, more communal society, such as the sort in which our brains evolved, confidence would be a much more accurate signal of knowledge and abilities." Ibid., 108.

CHAUFFEUR KNOWLEDGE

The story with Max Planck is probably invented: Charlie Munger, University of Southern California School of Law Commencement, May 13, 2007. Printed in Charles T. Munger, *Poor Charlie's Almanack*, expanded 3rd ed. (Virginia Beach, VA: The Donning Company Publishers, 2006), 399 and 435.

"You have to stick within what I call your circle of competence. You have to know what you understand and what you don't understand. It's not terribly important how big the circle is. But it is terribly important that you know where the perimeter is." In: Peter Bevelin, *Seeking Wisdom: From Darwin to Munger* (Malmö, Sweden: PCA Publications, 2007), 253.

"Again, that is a very, very powerful idea. Every person is going to have a circle of competence. And it's going to be very hard to enlarge that circle. If I had to make my living as a musician . . . I can't even think of a level low enough to describe where I would be sorted out to if music were the measuring standard of the civilization. So you have to figure out what your own aptitudes are. If you play games where other people have their aptitudes and you don't, you're going to lose. And that's as close to certain as any prediction that you can make. You have to figure out where you've got an edge. And you've got to play within your own circle of competence." Charlie Munger, "A Lesson on Elementary Worldly Wisdom as It Relates to Investment Management and Business," University of Southern California, 1994, in Munger, *Poor Charlie's Almanack*, 192.

"In the 2005 comedy-drama *The Weather Man*, the title character (played by Nicolas Cage) is paid well but receives little respect for his job, which consists entirely of acting authoritative while reading forecasts prepared by others." Christopher Chabris and Daniel Simons, *The Invisible Gorilla: And Other Ways Our Intuitions Deceive Us* (New York: Crown, 2010), 143.

ILLUSION OF CONTROL

The giraffe example from Max Gunther, *The Luck Factor: Why Some People Are Luckier Than Others and How You Can Become One of Them* (Petersfield, UK: Harriman House, 1977), chapter 3.

On rolling dice in casinos: J. M. Henslin, "Craps and Magic," *American Journal of Sociology* 73 (1967): 316–30.

Scott Plous, *The Psychology of Judgment and Decision Making* (New York: McGraw-Hill, 1993), 171.

The original study: Ellen J. Langer and J. Roth, "Heads I Win, Tails It's Chance: The Illusion of Control as a Function of the Sequence of Outcomes in a Purely Chance Task," *Journal of Personality and Social Psychology* 32, no. 6 (December 1975): 951–55.

Psychologist Roy Baumeister has shown that people tolerate more pain if they feel they understand their disease. The chronically ill cope much better when doctors can name the disease and explain what it is and does. It doesn't even have to be true. The effect works even if there is no proven cure for the disease. See: Roy F. Baumeister, *The Cultural Animal: Human Nature, Meaning, and Social Life* (Oxford, UK: Oxford University Press, 2005), 98–103.

People gain control by bringing the environment in line with their wishes (primary) but also by bringing their wishes in line with the environment (secondary). The illusion of control is part of the former strategies. This is the paper on it: Fred Rothbaum, John R. Weisz, and Samuel S. Snyder, "Changing the World and Changing the Self: A Two-Process Model of Perceived Control," *Journal of Personality and Social Psychology* 42, no. 1 (1982): 5–37.

The original experiment with two buttons: Herbert M. Jenkins and William C. Ward, "Judgment of Contingency between Responses and Outcomes," *Psychological Monographs* 79 (1965): 1–17.

The later experiment with just one button and no obligation to push the button. The subjects still had the illusion of control: Lorraine G. Allan and Herbert M. Jenkins, "The Judgment of Contingency and the Nature of the Response Alternatives," *Canadian Journal of Psychology* 34 (1980): 1–11.

The following four references shed light on placebo buttons:

Dan Lockton, "Placebo Buttons, False Affordances and Habit-Forming," *Design with Intent*, blog (http://architectures.danlockton.co.uk /2008/10/01/placebo-buttons-false-affordances-and-habit-forming/).

Michael Luo, "For Exercise in New York Futility, Push Button," *New York Times*, February 27, 2004.

Nick Paumgarten, "Up and Then Down—The Lives of Elevators," *New Yorker*, April 21, 2008.

Jared Sandberg, "Employees Only Think They Control Thermostat," *Wall Street Journal*, January 15, 2003.

INCENTIVE SUPER-RESPONSE TENDENCY

For an overview of Charlie Munger's thoughts on the *incentive super-response tendency*, read: Charles T. Munger, *Poor Charlie's Almanack*, expanded 3rd ed. (Virginia Beach, VA: The Donning Company Publishers, 2006), 450–57.

Charles T. Munger: "Perhaps the most important rule in management is: 'Get the incentives right.' " Ibid., 451.

The story with the fish lures: Ibid., 199.

REGRESSION TO MEAN

Beware: *Regression to mean* is not a causal correlation; it is purely statistical.

Daniel Kahneman: "I had the most satisfying Eureka experience of my career while attempting to teach flight instructors that praise is more effective than punishment for promoting skill-learning. When I had finished my enthusiastic speech, one of the most seasoned instructors in the audience raised his hand and made his own short speech, which began by conceding that positive reinforcement might be good for the birds, but went on to deny that it was optimal for flight cadets. He said, 'On many occasions I have praised flight cadets for clean execution of some aerobatic maneuver, and in general when they try it again, they do worse. On the other hand, I have often screamed at cadets for bad execution, and in general they do better the next time. So please don't tell us that reinforcement works and punishment does not, because the opposite is the case.' This was a joyous moment, in which I understood an important truth about the world."

Quote: Wikipedia entry, "Regression toward the Mean."

OUTCOME BIAS

The story with the monkeys, see: Burton Gordon Malkiel, *A Random Walk Down Wall Street: The Time-Tested Strategy for Successful Investing* (New York: W.W. Norton, 1973), 26.

Jonathan Baron and John C. Hershey, "Outcome Bias in Decision Evaluation," *Journal of Personality and Social Psychology* 54, no. 4 (1988): 569–79.

In case you want to calculate the example with the surgeons on your own, take any textbook on statistics and go to the chapter on urn models and "drawing with replacement." With no skills involved, the probabilities are as follows: nobody dies: 32.8 percent. One person dies: 41.1 percent. Two patients die: 20.5 percent. Three patients die: 5.1 percent. Four patients die: 0.6 percent. Five patients die: virtually zero probability.

See also: Nassim Nicholas Taleb, *Fooled by Randomness: The Hidden Role of Chance in Life and in the Markets*, 2nd updated ed. (New York: Random House, 2004), 154.

For the *historian error*, see also: David Hackett Fischer, *Historians' Fallacies: Toward a Logic of Historical Thought* (New York: Harper, 1970), 209–13.

PARADOX OF CHOICE

The Barry Schwartz video *The Paradox of Choice* can be found on TED.com.

Barry Schwartz, *The Paradox of Choice: Why More Is Less* (New York: Harper, 2004).

The problems with the *paradox of choice* are even more serious that those presented in the text. Tests have confirmed that decision making depletes energy that is later needed to keep emotional impulses in check. See: Roy F. Baumeister, *The Cultural Animal: Human Nature, Meaning, and Social Life* (Oxford, UK: Oxford University Press, 2005), 316–25.

People like autonomy but dislike making highly consequential decisions. See: Simona Botti, Kristina Orfali, and Sheena S. Iyengar, "Tragic Choices: Autonomy and Emotional Response to Medical Decisions," *Journal of Consumer Research* 36, no. 3 (2009): 337–52.

The more choice we have, the less satisfied we are after having made the choice. See: Sheena S. Iyengar, Rachael E. Wells, and Barry Schwartz, "Doing Better but Feeling Worse: Looking for the 'Best' Job Undermines Satisfaction," *Psychological Science* 17, no. 2 (2006): 143–50.

"Letting people think they have some choice in the matter is a powerful tool for securing compliance." Baumeister, *The Cultural Animal*, 323.

LIKING BIAS

Joe Girard, *How to Sell Anything to Anybody* (New York: Simon & Schuster, 1977).

"We rarely find that people have good sense unless they agree with us." (La Rochefoucauld)

Robert Cialdini dedicated an entire chapter to the *liking bias*: Robert B. Cialdini, *Influence: The Psychology of Persuasion*, rev. ed. (New York: HarperCollins, 1993), 167–207.

ENDOWMENT EFFECT

Dan Ariely, *Predictably Irrational: The Hidden Forces That Shape Our Decisions*, expanded ed. (New York: Harper Perennial, 2010), chapter 7, "The High Price of Ownership," 127–38.

The coffee mugs: Daniel Kahneman, Jack Knetsch, and Richard Thaler, "Experimental Test of the Endowment Effect and the Coase Theorem," *Journal of Political Economy* 98, no. 6 (1990): 1325–48.

Transactions don't happen if the lowest price a seller is ready to accept is higher than the highest price a seller is willing to pay. Why this often is the case: Ziv Carmon and Dan Ariely, "Focusing on the Forgone: How Value Can Appear So Different to Buyers and Sellers," *Journal of Consumer Research* 27 (2000): 360–70.

" . . . cutting your losses is a good idea, but investors hate to take losses because, tax considerations aside, a loss taken is an acknowledgment of error. Loss-aversion combined with ego leads investors to gamble by clinging to their mistakes in the fond hope that some day the market will vindicate their judgment and make them whole." Peter L. Bernstein, *Against the Gods: The Remarkable Story of Risk* (New York: Wiley, 1996), 276.

"A loss has about two and a half times the impact of a gain of the same magnitude." Niall Ferguson, *The Ascent of Money: A Financial History of the World* (New York: Penguin Press, 2008), 345.

"Losing ten dollars is perceived as a more extreme outcome than gaining ten dollars. In a sense, you know you will be more unhappy about losing ten dollars than you would be happy about winning the same amount, and so you refuse, even though a statistician or accountant would approve of taking the bet." Roy F. Baumeister, *The Cultural Animal: Human Nature, Meaning, and Social Life* (Oxford, UK: Oxford University Press, 2005), 319.

The more work you put into something, the more ownership you begin to feel for it (also called the "IKEA effect"). Michael I. Norton, Daniel Mochon, and Dan Ariely, "The 'IKEA Effect': When Labor Leads to Love" (working paper 11–091, Harvard Business School, March 2011).

COINCIDENCE

The story about the church explosion: Luke Nichols, "Church Explosion 60 Years Ago Not Forgotten," *Beatrice Daily Sun*, March 1, 2010.

Scott Plous, *The Psychology of Judgment and Decision Making* (New York: McGraw-Hill, 1993), 164.

For a good discussion on miracles, see: Peter Bevelin, *Seeking Wisdom: From Darwin to Munger* (Malmö, Sweden: PCA Publications, 2007), 167.

Numerous readers have contacted me regarding the story of the exploding church. They point out that the probability of all fifteen members arriving late is infinitesimally small. Let's assume, for example, that there is a 5 percent probability that a member will arrive thirty minutes late—meaning every twentieth rehearsal, or around twice a year, someone will come late, and that there is no correlation between individuals' late coming. This means the probability that all fifteen members will arrive late is

0.05 to the power of 15. This gives us a result of 3 times 10 to the power of -20. This calculation is correct, but imagine that the probabilities are correlated, which I believe is the case. How often does it happen that a drama or sports club has a terrible ambiance, so no one races to get to the next practice? In the very beginning of my literary career I had readings for which we'd sold thirty tickets, but not one person showed up. The weather was miserable and something more exciting was on television. In short (and without evidence for it), I believe that the probabilities were highly correlated. It certainly is the case with the married couple whose car didn't start.

Of course, the probability does not increase exactly by a factor of 100 if you have a hundred other friends. Imagine the probability is 2 percent that a friend calls just as you think about him. This does not become 200 percent if you have a hundred friends. Rather, it is $1-0.98^{100} = 86.7$ percent.

GROUPTHINK

Irving L. Janis, *Groupthink: Psychological Studies of Policy Decisions and Fiascoes*, 2nd ed. (Boston: Houghton Mifflin, 1982).

An opposite case of *groupthink* is *swarm intelligence* (James Surowiecki, *The Wisdom of Crowds* [New York: Doubleday, 2004]). Here is an overview: The large mass of average people (i.e., not a pool of experts) often finds remarkably correct solutions. Francis Galton (1907) demonstrated this in a nice experiment: He attended a cattle fair, which was also running a competition to guess the weight of an ox. Galton reckoned the visitors would not be up to the challenge and decided to statically evaluate the almost eight hundred guesses. The median of the estimates (1,197 pounds) was astonishingly close to the real weight of the ox (1,207 pounds). *Groupthink* occurs when participants interact. Swarm intelligence, on the other hand, occurs when players act independently of one another (e.g., when making guesses), which happens less and less. Swarm intelligence is very difficult to replicate scientifically.

NEGLECT OF PROBABILITY

Alan Monat, James R. Averill, and Richard S. Lazarus, "Anticipatory Stress and Coping Reactions under Various Conditions of Uncertainty," *Journal of Personality and Social Psychology* 24, no. 2 (November 1972): 237–53.

"Probabilities constitute a major human blind spot and hence a major focus for simplistic thought. Reality (especially social reality) is essentially probabilistic, but human thought prefers to treat it in simple, black-

and-white categories." Roy F. Baumeister, *The Cultural Animal: Human Nature, Meaning, and Social Life* (Oxford, UK: Oxford University Press, 2005), 206.

Since we have no intuitive understanding of probabilities, we also have no intuitive understanding of risk. Thus, stock market crashes must happen again and again to make hidden risks visible. It took an amazingly long time for economists to understand this. See: Peter L. Bernstein, *Against the Gods: The Remarkable Story of Risk* (New York: Wiley, 1996), 247–48.

However, what many economists and investors have not yet grasped is: Volatility is a poor measure of risk. And yet they use it in their evaluation models. See the following quote from Charlie Munger: "How can professors spread this nonsense that a stock's volatility is a measure of risk? I've been waiting for this craziness to end for decades. It's been dented, but it's still out there." Charles T. Munger, *Poor Charlie's Almanack*, expanded 3rd ed. (Virginia Beach, VA: The Donning Company Publishers, 2006), 101.

For a full discussion on how we (incorrectly) perceive risk: Paul Slovic, *The Perception of Risk* (London: Earthscan, 2000).

If the potential outcome of a technology is emotionally powerful, the risk (1 percent or 99 percent) has almost no baring on the attractiveness or unattractiveness of that technology. Paul Slovic, Melissa Finuane, Ellen Peters, and Donald G. MacGregor, "The Affect Heuristic," in Thomas Gilovich, Dale Griffin, and Daniel Kahneman (eds.), *Heuristics and Biases: The Psychology of Intuitive Judgment* (Cambridge, UK: Cambridge University Press, 2002), 409.

People are very sensitive to departures from absolute certainty and impossibility. But they are not very sensitive to departures from mid-range probabilities. See: Yuval Rottenstreich and Christopher K. Hsee, "Money, Kisses, and Electric Shocks: On the Affective Psychology of Risk," *Psychological Science* 12 (2001): 185–90.

An example is the Delaney Clause of the Food and Drug Act of 1958, which stipulated a total ban on synthetic carcinogenic food additives. The Delaney Clause stated, "No additive shall be deemed safe if it is found to induce cancer when ingested by man or animal."

SCARCITY ERROR

Robert B. Cialdini, *Influence: The Psychology of Persuasion*, rev. ed. (New York: HarperCollins, 1993), 237–71.

The cookie experiment, see: Stephen Worchel, Jerry Lee, and Akanabi

Adewole, "Effects of Supply and Demand on Ratings of Object Value," *Journal of Personality and Social Psychology* 32, no. 5 (November 1975): 906–14.

For the poster story, see: Roy F. Baumeister, *The Cultural Animal: Human Nature, Meaning, and Social Life* (Oxford, UK: Oxford University Press, 2005), 102.

The same works with music records instead of posters: Jack W. Brehm, Lloyd K. Stires, John Sensenig, and Janet Shaban, "The Attractiveness of an Eliminated Choice Alternative," *Journal of Experimental Social Psychology* 2, no. 3 (1966): 301–13.

Jack W. Brehm and Sharon S. Brehm frame the behavior as "reactance." Brehm and Brehm, *Psychological Reactance: A Theory of Freedom and Control* (New York: Academic Press, 1981).

BASE-RATE NEGLECT

The aphorism "When you hear hoofbeats behind you, don't expect to see a zebra" was coined in the late 1940s. Since horses are the most commonly encountered hoofed animal and zebras are very rare, logically you could confidently guess that the animal making the hoof beats is probably a horse. By 1960, the aphorism was widely known in medical circles. Source: http://en.wikipedia.org/wiki/Zebra_(medicine).

The example with the Mozart fan, see: Roy F. Baumeister, *The Cultural Animal: Human Nature, Meaning, and Social Life* (Oxford, UK: Oxford University Press, 2005), 206–7.

The classic study on the *base-rate neglect* is: Daniel Kahneman and Amos Tversky, "On the Psychology of Prediction," *Psychological Review* 80 (1973): 237–51.

The vignette with the wine tasting: Nassim Nicholas Taleb, personal communication and early manuscript of *The Black Swan*.

See also: Scott Plous, *The Psychology of Judgment and Decision Making* (New York: McGraw-Hill, 1993), 115–16.

GAMBLER'S FALLACY

One of the classic papers is: Iddo Gal and Jonathan Baron, "Understanding Repeated Simple Choices," *Thinking and Reasoning* 2, no. 1 (May 1, 1996): 81–98.

The *gambler's fallacy* is also called the "Monte Carlo fallacy." You can find the example from 1913 in the footnote of: Jonah Lehrer, *How We Decide* (New York: Houghton Mifflin Harcourt, 2009), 66.

The IQ example: Scott Plous, *The Psychology of Judgment and Decision Making* (New York: McGraw-Hill, 1993), 113.

See also: Thomas Gilovich, Robert Vallone, and Amos Tversky, "The Hot Hand in Basketball: On the Misperception of Random Sequences," in Thomas Gilovich, Dale Griffin, and Daniel Kahneman (eds.), *Heuristics and Biases: The Psychology of Intuitive Judgment* (Cambridge, UK: Cambridge University Press, 2002), 601–16.

The example with the loaded dice adapted from: Nassim Nicholas Taleb, *The Black Swan: The Impact of the Highly Improbable* (New York: Random House, 2007), 124.

THE ANCHOR

For the social security numbers and wheel of fortune, see: Dan Ariely, *Predictably Irrational: The Hidden Forces That Shape Our Decisions*, expanded ed. (New York: Harper Perennial, 2010), chapter 2. See also: Amos Tversky and Daniel Kahneman, "Judgment under Uncertainty: Heuristics and Biases," *Science* 185, no. 4157 (September 27, 1974): 1124–31.

The Abraham Lincoln example—albeit in modified form, see: Nicholas Epley and Thomas Gilovich, "Putting Adjustment Back in the Anchoring and Adjustment Heuristic," in Thomas Gilovich, Dale Griffin, and Daniel Kahneman (eds.), *Heuristics and Biases: The Psychology of Intuitive Judgment* (Cambridge, UK: Cambridge University Press, 2002), 139–49.

Also slightly modified in: Ulrich Frey and Johannes Frey, *Fallstricke; Die häufigsten Denkfehler in Alltag und Wissenschaft* (Munich: Beck, 2009), 40. There is no English translation of this book.

The Attila anecdote, see: Edward J. Russo and Paul. J. H. Shoemaker, *Decision Traps: The Ten Barriers to Decision-Making and How to Overcome Them* (New York: Simon & Schuster, 1989), 6.

On estimating house prices, see: Gregory B. Northcraft and Margaret A. Neale, "Experts, Amateurs, and Real Estate: An Anchoring-and-Adjustment Perspective on Property Pricing Decisions," *Organizational Behavior and Human Decision Processes* 39 (1987): 84–97.

Anchoring in negotiation and sales situations, see: Ilana Ritov, "Anchoring in Simulated Competitive Market Negotiation," *Organizational Behavior and Human Decision Processes* 67, no. 1 (July 1996): 16–25.

We all know the extraordinarily high requests for damages in liability lawsuits. One hundred million dollars for burning your fingers on a coffee cup. These requests work—thanks to anchoring. See: Gretchen B.

Chapman and Brian H. Bornstein, "The More You Ask For, the More You Get: Anchoring in Personal Injury Verdicts," *Applied Cognitive Psychology* 10 (1996): 519–40.

INDUCTION

The goose example comes from Nassim Taleb, though he used a Thanksgiving turkey. Taleb borrowed the example from Bertrand Russell (he used a chicken), who, in turn, borrowed it from David Hume. See: Nassim Nicholas Taleb, *The Black Swan: The Impact of the Highly Improbable* (New York: Random House, 2007), 40.

Induction is a major topic in epistemology: How can we make statements about the future when the past is all we have? Answer: We cannot. Each case of induction is always fraught with uncertainty. The same goes for causality: We can never know if things are causally linked, even if we have observed them a million times. David Hume covered these issues brilliantly in the eighteenth century. Later it was Karl Popper who warned against our naive belief in induction.

LOSS AVERSION

The original research that brought the *loss aversion* to light stems from Daniel Kahneman and Amos Tversky. They called their findings Prospect Theory for lack of a better word. This is the original paper: Daniel Kahneman and Amos Tversky, "Prospect Theory: An Analysis of Decision under Risk," *Econometrica* 47, no. 2 (1979): 263–92. This paper generated an avalanche of follow-up research, mostly confirming the original findings.

The example with the breast-cancer awareness campaign, see: Beth E. Meyerowitz and Shelly Chaiken, "The Effect of Message Framing on Breast Self-Examination Attitudes, Intentions, and Behavior," *Journal of Personality and Social Psychology* 52, no. 3 (March 1987): 500–510. The emphasis in the quoted text is mine. The study included two more short paragraphs with a gain-frame or loss-frame, respectively.

Recent studies, however, don't see such a clear results. See: Daniel J. O'Keefe and Jakob D. Jensen, "The Relative Persuasiveness of Gain-Framed and Loss-Framed Messages for Encouraging Disease Prevention Behaviors: A Meta-Analytic Review," *Journal of Health Communication*, 12, no. 7 (2007): 623–44, DOI: 10.1080/10810730701615198.

We react more strongly to negative enticements than to positive ones. See: Roy F. Baumeister, *The Cultural Animal: Human Nature, Meaning, and Social Life* (Oxford, UK: Oxford University Press, 2005), 318–21.

This research paper explains that we're not the only species prone to *loss aversion*. Monkeys also fall for it, albeit for other reasons: A. Silberberg et al., "On Loss Aversion in Capuchin Monkeys," *Journal of the Experimental Analysis of Behavior* 89 (2008): 145–55.

SOCIAL LOAFING

David A. Kravitz and Barbara Martin, "Ringelmann Rediscovered: The Original Article," *Journal of Personality and Social Psychology* 50, no. 5 (1986): 936–41.

Bibb Latané, Kippling Williams, and Stephen Harkins, "Many Hands Make Light the Work: The Causes and Consequences of Social Loafing," *Journal of Personality and Social Psychology* 37, no. 6 (1979): 822–32.

See also: Scott Plous, *The Psychology of Judgment and Decision Making* (New York: McGraw-Hill, 1993), 192–93.

To learn more about risky shift, see: Dean G. Pruitt, "Choice Shifts in Group Discussion: An Introductory Review," *Journal of Personality and Social Psychology* 20, no. 3 (1971): 339–60, and Serge Moscovici and Marisa Zavalloni, "The Group as a Polarizer of Attitudes," *Journal of Personality and Social Psychology* 12, no. 2 (1969): 125–35.

EXPONENTIAL GROWTH

Where does the number 70 come from? It is the natural logarithm of 2 times 100. That's 69.3, which is close enough to 70. If you'd be interested in the tripling time, you can use the natural logarithm of 3. If you'd be interested in the quintupling time, you'd use the natural logarithm of 5.

For good examples of *exponential growth*, see: Dietrich Dörner, *Die Logik des Misslingens: Strategisches Denken in komplexen Situationen* (Reinbek, Germany: Rororo Publisher, 2003), 161–71. There is no English translation of this book.

See also: Hans-Hermann Dubben and Hans-Peter Beck-Bornholdt, *Der Hund, der Eier legt: Erkennen von Fehlinformation durch Querdenken* (Reinbek, Germany: Rororo Publisher, 2006), 120. There is no English translation of this book.

Exponential population growth was a hot topic during the 1970s when resource scarcity came to the fore. See: Donella H. Meadows, Dennis L. Meadows, Jorgen Randers, and William W. Behrens III, *The Limits to Growth* (New York: Universe Books, 1972). The "new economy," which set the stage for the "great moderation" and promoted growth free from inflation and such scarcity, cleared the issue from the table. However, since the raw material shortages of 2007, we know that this continues to be a problem—especially since the global population is still growing exponentially.

WINNER'S CURSE

The classic source: Richard H. Thaler, "The Winner's Curse," *Journal of Economic Perspectives* 2, no. 1 (Winter 1988): 191–202.

If you need to outdo another person, see: Deepak Malhotra, "The Desire to Win: The Effects of Competitive Arousal on Motivation and Behavior," *Organizational Behavior and Human Decision Processes* 111, no. 2 (March 2010): 139–46.

There are numerous examples of the *winner's curse* in action. For example, in book publishing. "The problem is, simply, that most of the auctioned books are not earning their advances. In fact, very often such books have turned out to be dismal failures whose value was more perceived than real." John P. Dessauer, *Book Publishing* (New York: Bowker, 1981), 33. I sincerely hope that the book you hold in your hands is an exception.

How much would you pay for $100? An example from Scott Plous, *The Psychology of Judgment and Decision Making* (New York: McGraw-Hill, 1993), 248–49. Plous describes it with $1 instead of $100. The mechanics are the same.

"The Warren Buffett rule for open-outcry auctions: Don't go." Charles T. Munger, *Poor Charlie's Almanack*, expanded 3rd ed. (Virginia Beach, VA: The Donning Company Publishers, 2006), 494.

Value destroying M&A, in: Werner Rehm, Robert Uhlaner, and Andy West, "Taking a Longer-Term Look at M&A Value Creation," *McKinsey on Finance* 42 (Winter 2012): 8.

FUNDAMENTAL ATTRIBUTION ERROR

Stanford psychologist Lee Ross described this for the first time, see: Lee Ross, "The Intuitive Psychologist and His Shortcomings: Distortions in the Attribution Process," in L. Berkowitz (ed.), *Advances in Experimental Social Psychology*, vol. 10 (New York: Academic Press, 1977).

The experiment with the speech, see: Edward E. Jones and Victor A. Harris, "The Attribution of Attitudes," *Journal of Experimental Social Psychology* 3 (1967): 1–24. Actually, there are three experiments in that paper, two about Fidel Castro, one about racial segregation in the United States. The point of interest here is the result after the first Fidel Castro experiment: "Perhaps the most striking result of the first experiment was the tendency to attribute correspondence between behavior and private attitude even when the direction of the essay was assigned." Ibid., 7.

See also: Scott Plous, *The Psychology of Judgment and Decision Making* (New York: McGraw-Hill, 1993), 180–81.

Buffett: "A wise friend told me long ago, 'If you want to get a reputation as a good businessman, be sure to get into a good business.'" In: Berkshire Hathaway Inc. 2006 Annual Report, 11.

FALSE CAUSALITY

Hans-Hermann Dubben and Hans-Peter Beck-Bornholdt, *Der Hund, der Eier legt: Erkennen von Fehlinformation durch Querdenken* (Reinbek, Germany: Rororo Publisher, 2006), 175–78. Unfortunately, there is no English translation of this book.

The nice example using the stork. Ibid., 181.

Having books at home, see: "To Read or Not to Read: A Question of National Consequence," National Endowment for the Arts, Research Report #47, November 2007.

HALO EFFECT

The ultimate book about the *halo effect* in business, including the Cisco example: Phil Rosenzweig, *The Halo Effect—and the Eight Other Business Delusions That Deceive Managers* (New York: Free Press, 2007).

Thorndike defined the *halo effect* as "a problem that arises in data collection when there is carry-over from one judgment to another." Edward L. Thorndike, "A Constant Error on Psychological Rating," *Journal of Applied Psychology* 4 (1920): 25–29.

Richard E. Nisbett and Timothy D. Wilson, "The Halo Effect: Evidence for Unconscious Alteration of Judgments," *Journal of Personality and Social Psychology* 35, no. 4 (1977): 250–56.

ALTERNATIVE PATHS

The Russian roulette example: Nassim Nicholas Taleb, *Fooled by Randomness: The Hidden Role of Chance in Life and in the Markets*, 2nd updated ed. (New York: Random House, 2004), 23.

"It is hard to think of Alexander the Great or Julius Caesar as men who won only in the visible history, but who could have suffered defeat in others. If we have heard of them, it is simply because they took considerable risks, along with thousands of others, and happened to win. They were intelligent, courageous, noble (at times), had the highest possible obtainable culture in their day—but so did thousands of others who live in the musty footnotes of history." Ibid., 34.

"My argument is that I can find you a security somewhere among the 40,000 available that went up twice that amount every year without fail. Should we put the social security money into it?" Ibid.,146.

FORECAST ILLUSION

The classic book on the *forecast illusion* is: Philip E. Tetlock, *Expert Political Judgment: How Good Is It? How Can We Know?* (Princeton, NJ: Princeton University Press, 2005).

For a short summary: Philip E. Tetlock, "How Accurate Are Your Pet Pundits?," Project Syndicate/Institute for Human Sciences, 2006, accessed October 20, 2012. http://www.project-syndicate.org/commentary/how-accurate-are-your-pet-pundits.

Derek J. Koehler, Lyle Brenner, and Dale Griffin, "The Calibration of Expert Judgment: Heuristics and Biases Beyond the Laboratory," in Thomas Gilovich, Dale Griffin, and Daniel Kahneman (eds.), *Heuristics and Biases: The Psychology of Intuitive Judgment* (Cambridge, UK: Cambridge University Press, 2002), 686–715.

"The only function of economic forecasting is to make astrology look respectable." John Kenneth Galbraith quoted in *U.S. News & World Report*, March 7, 1988, 64.

The forecast anecdote from Tony Blair: Roger Buehler, Dale Griffin, and Michael Ross, "Inside the Planning Fallacy: The Causes and Consequences of Optimistic Time Predictions," in Gilovich, Griffin, and Kahneman (eds.), *Heuristics and Biases*, 270.

"There have been as many plagues as wars in history, yet always plagues and wars take people equally by surprise." Albert Camus, *The Plague*, part 1.

"I don't read economic forecasts. I don't read the funny papers." Warren Buffett quoted in "Buffett Builds Up Stake in UK Blue Chip," *Independent*, April 13, 1999, http://www.independent.co.uk/news/business/buffett-builds-up-stake-in-uk-blue-chip-1086992.html.

Harvard Professor Theodore Levitt: "It's easy to be a prophet. You make twenty-five predictions and the ones that come true are the ones you talk about." In: Peter Bevelin, *Seeking Wisdom: From Darwin to Munger* (Malmö, Sweden: PCA Publications, 2007), 167.

"There are 60,000 economists in the U.S., many of them employed full-time trying to forecast recessions and interest rates, and if they could do it successfully twice in a row, they'd all be millionaires by now. They'd have retired to Bimini where they could drink rum and fish for marlin. But as

far as I know, most of them are still gainfully employed, which ought to
tell us something." In: Peter Lynch, *One Up on Wall Street: How to Use
What You Already Know to Make Money in the Market* (New York: Simon
& Schuster, 2000), 85.

And since it is so pithy, here's another quote from the same book: "Thousands
of experts study overbought indicators, oversold indicators, head-and-
shoulder patterns, put-call ratios, the Fed's policy on money supply, foreign investment, the movement of the constellations through the heavens,
and the moss on oak trees, and they can't predict markets with any useful
consistency, any more than the gizzard squeezers could tell the Roman
emperors when the Huns would attack." Ibid.

Stock market analysts are especially good at retrospective forecasting: "The
analysts and the brokers. They don't know anything. Why do they
always downgrade stocks after the bad earnings come out? Where's the
guy that downgrades them before the bad earnings come out? That's
the smart guy. But I don't know any of them. They're rare, they're very
rare. They're rarer than Jesse Jackson at a Klan meeting." Marc Perkins
interviewed by Brett D. Fromson, The TSC Streetside Chat, part 2,
TheStreet.com, September 8, 2000.

Buffett: "When they make these offerings, investment bankers display their
humorous side: They dispense income and balance sheet projections
extending five or more years into the future for companies they barely had
heard of a few months earlier. If you are shown such schedules, I suggest that you join in the fun: Ask the investment banker for the *one-year*
budgets that his own firm prepared as the last few years began and then
compare these with what actually happened." In: Berkshire Hathaway,
Inc., letter to shareholders, 1989.

Warren Buffett: "I have no use whatsoever for projections or forecasts. They
create an illusion of apparent precision. The more meticulous they are,
the more concerned you should be. We never look at projections, but we
care very much about, and look very deeply at, track records." Berkshire Hathaway annual meeting, 1995, quoted in Andrew Kilpatrick, *Of
Permanent Value: The Story of Warren Buffett* (Birmingham, AL: AKPE,
2010), 1074.

Here is another great study that shows the inability for experts to forecast.
Gustav Torngren and Henry Montgomery asked participants to select
the stock from a pair of stocks that would outperform each month. They
were known blue chip names, and the players were given the prior twelve
months' performance for each stock. Participants included lay people
(undergrads in psychology) and professional investors. Both groups

performed worse than sheer luck. Both would have fared better by tossing a coin. Overall, the laypeople were 59 percent confident in their stock picking abilities, the experts 65 percent. See: Gustav Torngren and Henry Montgomery, "Worse Than Chance? Performance and Confidence among Professionals and Laypeople in the Stock Market," *Journal of Behavioural Finance* 5, no. 3 (2004): 148–53.

CONJUNCTION FALLACY

The Chris story is a modified version of the so-called Bill story and Linda story by Tversky and Kahneman: Amos Tversky and Daniel Kahneman, "Extension versus Intuitive Reasoning: The Conjunction Fallacy in Probability Judgment," *Psychological Review* 90, no. 4 (October 1983): 293–315. Thus, the *conjunction fallacy* is often referred to as the "Linda problem."

The example using oil consumption: Ibid., 308. Another interesting example of the *conjunction fallacy* can be found in the same paper. What is more probable? (a) "a complete suspensions of diplomatic relations between the US and the Soviet Union, sometime in 1983," or (b) "a Russian invasion of Poland, and a complete suspensions of diplomatic relations between the US and the Soviet Union, sometime in 1983." Many more people opted for the more plausible scenario B, although it is less likely.

On the two types of thinking—intuitive versus rational, or system 1 versus system 2, see: Daniel Kahneman, "A Perspective on Judgment and Choice: Mapping Bounded Rationality," *American Psychologist* 58 (September 2003): 697–720. Or you can read Kahneman's *Thinking, Fast and Slow* (New York: Farrar, Straus and Giroux, 2011), which is all about system 1 versus system 2.

A much simpler version of the *conjunction fallacy* is the following question that has been posed to children: "In summer at the beach are there more women or more tanned women?" Most children fell for (the more representative or available) "tanned women." See: Franca Agnoli, "Development of Judgmental Heuristics and Logical Reasoning: Training Counteracts the Representativeness Heuristic," *Cognitive Development* 6, no. 2 (April–June 1991): 195–217.

Tversky and Kahneman asked: What is more likely, that a seven-letter word randomly selected from a novel would end in *ing* or has the letter "n" as its sixth letter? This highlights both the availability bias and the conjunction fallacy. All seven-letter words ending with *ing* have the letter "n" as its sixth letter, but not all with the letter "n" as its sixth letter end in *ing*. Again, the driving force for the *conjunction fallacy* is the availability bias. Words ending with *ing* come to mind more easily. See: Tversky and

Kahneman, "Extensional versus Intuitive Reasoning: The Conjunction Fallacy in Probability Judgment," 295.

The story with the terrorism insurance is adapted from Nassim Nicholas Taleb, *The Black Swan: The Impact of the Highly Improbable* (New York: Random House, 2007), 76–77.

FRAMING

Amos Tversky and Daniel Kahneman, "The Framing of Decisions and the Psychology of Choice," *Science* 211, no. 4481 (January 30, 1981): 453–58.

The *framing* effect in medicine, see: Robyn M. Dawes, *Everyday Irrationality: How Pseudo-Scientists, Lunatics, and the Rest of Us Systematically Fail to Think Rationally* (New York: Westview Press, 2001), 3–8.

R. Shepherd, P. Sparks, S. Bellier, and M. M. Raats, "The Effects of Information on Sensory Ratings and Preferences: The Importance of Attitudes," *Food Quality and Preference* 3, no. 3 (1992): 147–55.

ACTION BIAS

Michael Bar-Eli, Ofer H. Azar, Ilana Ritov, Yael Keidar-Levin, and Galit Schein, "Action Bias among Elite Soccer Goalkeepers: The Case of Penalty Kicks," *Journal of Economic Psychology* 28, no. 5 (2007): 606–21.

The quote from Charlie Munger: "We've got great flexibility and a certain discipline in terms of not doing some foolish thing just to be active— discipline in avoiding just doing any damn thing just because you can't stand inactivity." In: Wesco Financial annual meeting, 2000, *Outstanding Investor Digest*, December 18, 2000, 60.

Warren Buffett successfully avoids the *action bias*: "We don't get paid for *activity*, just for being *right*. As to how long we'll *wait*, we'll wait *indefinitely*." Warren Buffett, 1998 Berkshire Hathaway annual meeting.

"The stock market is a no-called-strike game. You don't have to swing at everything—you can wait for your pitch. The problem when you're a money manager is that your fans keep yelling, 'Swing, you bum!' " Warren Buffett, 1999 Berkshire Hathaway annual meeting.

"It takes character to sit there with all that cash and do nothing. I didn't get to where I am by going after mediocre opportunities." Charlie Munger, *Poor Charlie's Almanack*, expanded 3rd ed. (Virginia Beach, VA: The Donning Company Publishers, 2006), 61.

"Charlie realizes that it is difficult to find something that is really good. So, if you say 'No' ninety percent of the time, you're not missing much in the world." Otis Booth in ibid., 99.

Charlie Munger: "There are huge advantages for an individual to get into a position where you make a few great investments and just sit on your ass: You're paying less to brokers. You're listening to less nonsense." Ibid., 209.

The example with the police officers in: "Action Bias in Decision Making and Problem Solving," *Ambiguity Advantage*, blog, February 21, 2008.

OMISSION BIAS

Jonathan Baron, *Thinking and Deciding* (Cambridge, UK: Cambridge University Press, 2000), 407–8 and 514.

To get around the omission bias, put yourself in the shoes of the harmed individual. If you were that baby about to get vaccinated, what is more preferable to you: a 10/10,000 chance of death from the disease or a 5/10,000 chance death from the vaccine? And does it matter if these chances are a matter of commission or omission? Ibid., 407.

D. A. Asch, Jonathan Baron, J . C. Hershey, H. Kunreuther, J. R. Meszaros, Ilana Ritov, and M. Spranca, "Omission Bias and Pertussis Vaccination," *Medical Decision Making* 14, no. 2 (April–June 1994): 118–23.

There is some confusion as to whether a behavior is due to the *omission bias*, the status quo bias, or social norm. Baron and Ritov disentangle these questions in this paper: Jonathan Baron and Ilana Ritov, "Omission Bias, Individual Differences, and Normality," *Organizational Behavior and Human Decision Processes* 94 (2004): 74–85.

The following paper deals with the *omission bias* in legal practice in Switzerland. It is only available in German: Mark Schweizer, "Der Unterlassungseffekt," chapter from "Kognitive Täuschungen vor Gericht" (PhD dissertation, University of Zurich, 2005), 108–23.

SELF-SERVING BIAS

Just as in the "taking out the garbage" example, Ross and Sicoly asked husbands and wives to which percentage they are responsible for activities like cleaning the house, making breakfast, causing arguments. Each spouse overestimated his or her role. The answers always added up to more than 100 percent. Read: Ross and Sicoly, "Egocentric Bias in Availability and Attribution."

Barry R. Schlenker and Rowland S. Miller, "Egocentrism in Groups: Self-Serving Biases or Logical Information Processing?," *Journal of Personality and Social Psychology* 35, no. 10 (October 1977): 755–64.

The following research modifies that view that we always attribute failure to outside factors: Dale T. Miller and Michael Ross, "Self-Serving Biases in the Attribution of Causality: Fact or Fiction?," *Psychological Bulletin* 82 (1975): 213–25.

Roy F. Baumeister, *The Cultural Animal: Human Nature, Meaning, and Social Life* (Oxford, UK: Oxford University Press, 2005), 214–19.

"Of course you also want to get the self-serving bias out of your mental routines. Thinking that what's good for you is good for the wider civilization, and rationalizing foolish or evil conduct, based on your subconscious tendency to serve yourself, is a terrible way to think." Charles T. Munger: *Poor Charlie's Almanack*, expanded 3rd ed. (Virginia Beach, VA: The Donning Company Publishers, 2006), 432.

Joel T. Johnson, Lorraine M. Cain, Toni L. Falke, Jon Hayman, and Edward Perillo, "The 'Barnum Effect' Revisited: Cognitive and Motivational Factors in the Acceptance of Personality Descriptions," *Journal of Personality and Social Psychology* 49, no. 5 (November 1985): 1378–91.

This is an example of a study with school grades: Robert M. Arkin and Geoffrey M. Maruyama, "Attribution, Affect and College Exam Performance," *Journal of Educational Psychology* 71, no. 1 (February 1979): 85–93.

See this video on grades on TED.com: Dan Ariely, *Why We Think It's OK to Cheat and Steal (Sometimes)*.

The *self-serving bias* is sometimes also called "egocentric bias." Sometimes, the scientific literature differentiates between the two, especially when it comes to group settings. The *self-serving bias* claims credit for positive outcomes only. The egocentric bias, however, claims credit even for negative outcomes. It is suggested that the egocentric bias is simply an availability bias in disguise because your own actions and contributions are more available to you (in memory) than the actions and contributions of the other group members. See: Ross and Sicoly, "Egocentric Biases in Availability and Attribution."

HEDONIC TREADMILL

The classic paper on the *hedonic treadmill* effect: Philip Brickman and D. T. Campbell, "Hedonic Relativism and Planning the Good Society," in M. H. Appley (ed.), *Adaptation-Level Theory: A Symposium* (New York: Academic Press, 1971), 278–301. It focuses not just on income, but on improvements of consumer electronic and gadgets. We quickly adjust to the latest gadgets and their "happiness effect" fades away quickly.

Daniel T. Gilbert et al., "Immune Neglect: A Source of Durability Bias in

Affective Forecasting," *Journal of Personality and Social Psychology* 75, no. 3 (1989): 617–38.

Daniel T. Gilbert and Jane E. Ebert, "Decisions and Revisions: The Affective Forecasting of Changeable Outcomes," *Journal of Personality and Social Psychology* 82, no. 4 (2002): 503–14.

Daniel T. Gilbert, *Stumbling on Happiness* (New York: Alfred A. Knopf, 2006).

Major live dramas have almost no long-term impact on happiness. Daniel T. Gilbert, *Why Are We Happy?*, video on TED.com (http://www.youtube.com/watch?v=LTO_dZUvbJA).

Nassim Nicholas Taleb, *The Black Swan: The Impact of the Highly Improbable* (New York: Random House, 2007), 91.

Bruno S. Frey and Alois Stutzer, *Happiness and Economics: How the Economy and Institutions Affect Human Well-Being* (Princeton, NJ: Princeton University Press, 2002).

Subjective well-being (happiness) seems to be heavily influenced by genetics. In other words, it's chance! Socioeconomic status, educational attainment, family income, marital status, or religious commitment can account for no more than about 3 percent of the variance in subjective well-being. See: David Lykken and Auke Tellegen, "Happiness Is a Stochastic Phenomenon," *Psychological Science* 7, no. 3 (May 1996): 186–89.

Life satisfaction seems to be extremely stable over time, although it can be more volatile in the short term. See: Frank Fujita and Ed Diener, "Life Satisfaction Set Point: Stability and Change," *Journal of Psychology and Social Psychology* 88, no. 1 (2005): 158–64.

In case you are looking for more research on the topic: *hedonic treadmill* is also called "hedonic adaptation."

SELF-SELECTION BIAS

On incubation of funds: "A more deliberate form of self selection bias often occurs in measuring the performance of investment managers. Typically, a number of funds are set up that are initially incubated: kept closed to the public until they have a track record. Those that are successful are marketed to the public, while those that are not successful remain in incubation until they are. In addition, persistently unsuccessful funds (whether in an incubator or not) are often closed, creating survivorship bias. This is all the more effective because of the tendency of investors to pick funds from the top of the league tables regardless of the performance of the manager's other funds." Quoted from *Moneyterms*, http://moneyterms.co.uk/self-selection-bias/.

"It is not uncommon for someone watching a tennis game on television to be bombarded by advertisements for funds that did (until that minute) outperform other by some percentage over some period. But, again, why would anybody advertise if he didn't happen to outperform the market? There is a high probability of the investment coming to you if its success is caused entirely by randomness. This phenomenon is what economists and insurance people call adverse selection." Nassim Nicholas Taleb, *Fooled by Randomness: The Hidden Role of Chance in Life and in the Markets*, 2nd updated ed. (New York: Random House, 2004), 158.

ASSOCIATION BIAS

The story with the gas leak, see: Roy F. Baumeister, *The Cultural Animal: Human Nature, Meaning, and Social Life* (Oxford, UK: Oxford University Press, 2005), 280.

Buffett wants to hear the bad news—in plain terms. "Always tell us the bad news promptly. It is only the good news that can wait." In: Charles T. Munger, *Poor Charlie's Almanack*, expanded 3rd ed. (Virginia Beach, VA: The Donning Company Publishers, 2006), 472.

"Don't shoot the messenger" appears in Shakespeare's *Henry IV*, last act.

In the eighteenth century, many states, including the states in New England, employed town criers. Their task was to disseminate news—often bad news—for example, tax increases. In order to beat the "kill the messenger" syndrome, the states adopted a law (probably read aloud by the town crier), whereby injury or abuse of the crier earned the harshest penalty. Today we are no longer as civilized. We try to lock up the loudest criers. Such an example is Julian Assange, founder of Wikileaks.

BEGINNER'S LUCK

Nassim Nicholas Taleb, *The Black Swan: The Impact of the Highly Improbable* (New York: Random House, 2007), 109.

COGNITIVE DISSONANCE

Scott Plous, *The Psychology of Judgment and Decision Making* (New York: McGraw-Hill, 1993), 22–25.

The classic paper on *cognitive dissonance*: Leon Festinger and James M. Carlsmith, "Cognitive Consequences of Forced Compliance," *Journal of Abnormal and Social Psychology* 58 (1959): 203–10.

There is a French version of the sour-grapes rationalization: The fox wrongly

believes the grapes to be green instead of vermillion and sweet. See: Jon
Elster, *Sour Grapes: Studies in the Subversion of Rationality* (Cambridge,
UK: Cambridge University Press, 1983), 123–24.

One of investor George Soros's strengths, according to Taleb, is his complete
lack of *cognitive dissonance*. Soros can change his mind from one second
to the next—without the slightest sense of embarrassment. See: Nassim
Nicholas Taleb, *Fooled by Randomness: The Hidden Role of Chance in Life
and in the Markets*, 2nd updated ed. (New York: Random House, 2004),
239.

HYPERBOLIC DISCOUNTING

A range of research papers cover this topic. This is the first: Richard H.
Thaler, "Some Empirical Evidence on Dynamic Inconsistency," *Economic
Letters* 8 (1981): 201–7.

For the marshmallow test, see: Yuichi Shoda, Walter Mischel, and Philip K.
Peake, "Predicting Adolescent Cognitive and Self-Regulatory Compe-
tencies from Preschool Delay of Gratification: Identifying Diagnostic
Conditions," *Developmental Psychology* 26, no. 6 (1990): 978–86.

" . . . the ability to delay gratification is very adaptive and rational, but sometimes it fails and people grab for immediate satisfaction. The effect of the
immediacy resembles the certainty effect: People prefer the immediate
gain just as they prefer the guaranteed gain. And both of these suggest
that underneath the sophisticated thinking process of the cultural animal
there still lurk the simpler needs and inclinations of the social animal.
Sometimes these win out." Roy F. Baumeister, *The Cultural Animal: Human Nature, Meaning, and Social Life* (Oxford, UK: Oxford University
Press, 2005), 321.

What about very long periods of time? Suppose you run a restaurant and a
diner makes the following suggestion: Instead of paying his check of $100
today, he will pay you $1,700 in thirty years' time—that's a nice inter-
est rate of 10 percent. Would you go for it? Probably not. Who knows
what will happen in the next thirty years? So have you just committed a
thinking error? No. In contrast to *hyperbolic discounting*, higher interest
rates over long periods of time are quite advisable. In Switzerland (before
Fukushima), there was debate about a plan to build a nuclear power plant
with a payback period of thirty years. An idiotic idea. Who knows what
new technologies will come on the market during those thirty years? A
payback period of ten years would be justified, but not thirty years—and
that's not even mentioning the risks.

"BECAUSE" JUSTIFICATION

The Xerox experiment by Ellen Langer cited in Robert B. Cialdini, *Influence: The Psychology of Persuasion*, rev. ed. (New York: HarperCollins, 1993), 4.

The *"because" justification* works beautifully as long as the stakes are small (making copies). As soon as the stakes are high, people mostly listen attentively to the arguments. Noah Goldstein, Steve Martin, and Robert Cialdini, *Yes!—50 Scientifically Proven Ways to Be Persuasive* (New York: Free Press, 2008), 150–53.

DECISION FATIGUE

"The problem of decision fatigue affects everything from the careers of CEOs to the prison sentences of felons appearing before weary judges. It influences the behavior of everyone, executive and nonexecutive, every day." Roy Baumeister and John Tierney, *Willpower: Rediscovering the Greatest Human Strength* (New York: Penguin Press, 2011), 90.

The student experiment with the "deciders" and "non-deciders": Ibid., 91, 92.

The example with the judges: Ibid., 96–99.

The detailed paper on the judges' decisions: Shai Danziger, Jonathan Levav, and Liora Avnaim-Pesso, "Extraneous Factors in Judicial Decisions," *Proceedings of the National Academy of Science* 108, no. 17 (February 25, 2011): 6889–92.

Roy Baumeister, "Ego Depletion and Self-Control Failure: An Energy Model of the Self's Executive Function," *Self and Identity* 1, no. 2 (April 1, 2002): 129–36.

Kathleen D. Vohs, Roy F. Baumeister, Jean M. Twenge, Brandon J. Schmeichel, Dianne M. Tice, and Jennifer Crocker, "Decision Fatigue Exhausts Self-Regulatory Resources—But So Does Accommodating to Unchosen Alternatives," *Working paper*, 2005.

George Loewenstein, Daniel Read, and Roy Baumeister, *Time and Decision: Economic and Psychological Perspectives on Intertemporal Choice* (New York: Russell Sage Foundation, 2003), 208.

After the hard slog through the supermarket, consumers suffer *decision fatigue*. Retailers capitalize on this and place impulse buys, such as gum and candy, right next to cashiers—just before the finishing line of the decision marathon. See: John Tierney, "Do You Suffer from Decision Fatigue?," *New York Times Magazine*, August 17, 2011.

When to present it to your CEO? The best time is eight a.m. The CEO will be relaxed after a good night's sleep, and after breakfast his blood sugar level will be high—all perfect for making courageous decisions.

CONTAGION BIAS

Contagion bias is also called the "contagion heuristic."

The one-line summary of the *contagion bias*: "Once in contact, always in contact."

Thomas Gilovich, Dale Griffin, and Daniel Kahneman (eds.), *Heuristics and Biases: The Psychology of Intuitive Judgment* (Cambridge, UK: Cambridge University Press, 2002), 212.

See also the Wikipedia entry for the "Peace and Truce of God," accessed October 21, 2012.

Philip Daileader, *The High Middle Ages* (Chantilly, VA: The Teaching Company, 2001), course no. 869, lecture 3, beginning at ~26:30.

The example with the arrows comes from Kennedy vs. Hitler in: Gilovich, Griffin, and Kahneman (eds.), *Heuristics and Biases*, 205. The authors of the article (Paul Rozin and Carol Nemeroff) are not talking about "contagion" but about the "law of similarity." I have added the example of contagion heuristic, which in the broader sense deals with a penchant for magic.

Photos of mothers: A control group that did not use photos was better at hitting the targets. Participants behaved as if the photos contained magic powers that might hurt the real subjects. In a similar experiment, photographs of either John F. Kennedy or Hitler were pasted onto the targets. Although all students were trying to shoot as accurately as possible, those who had JFK in their crosshairs fared much worse. (Ibid.)

We do not like to move into recently deceased people's houses, apartments, or rooms. Conversely, companies love when their new offices previously housed successful companies. For example, when milo.com moved into 165 University Avenue in Palo Alto, there was a lot of press because Logitech, Google, and PayPal all used to be in that building. As if some "good vibes" would lift the start-ups in that building. It certainly has more to do with the proximity to Stanford University.

To calculate the number of molecules per breath: The atmosphere consists of approximately 10^{44} molecules. The total atmospheric mass is 5.1×10^{18} kg. Air density at sea level is about 1.2 kg/m^3. According to the Avogadro constant, there are 2.7×10^{25} molecules in a cubic meter of air. So, in one liter there are 2.7×10^{22} molecules. On average, we breathe about seven liters of air per minute (about one liter per breath) or 3,700 cubic meters per year. Saddam Hussein "consumed" 260,000 cubic meters of air in his life. Assuming he re-inhaled approximately 10 percent of that, we have 230,000 cubic meters of "Saddam-contaminated" air in the atmosphere. Thus 6.2×10^{30} molecules passed through Saddam's lungs, which are now scattered in the atmosphere. The concentration of these molecules in the atmosphere equals 6.2×10^{-14}. That makes 1.7 billion "Saddam-contaminated" molecules per breath.

See also: Carol Nemeroff and Paul Rozin, "The Makings of the Magical Mind: The Nature of Function of Sympathetic Magic," in Karl S. Rosengren, Carl N. Johnson, and Paul L. Harris (eds.), *Imagining the Impossible: Magical, Scientific, and Religious Thinking in Children* (Cambridge, UK: Cambridge University Press, 2000), 1–34.

THE PROBLEM WITH AVERAGES

Don't cross a river if it is (on average) four feet deep: Nassim Nicholas Taleb, *The Black Swan: The Impact of the Highly Improbable* (New York: Random House, 2007), 160.

The overall median wealth per family in the United States was $109,500 in 2007. See: Wikipedia Entry on "Wealth in the United States," accessed October 25, 2012, http://en.wikipedia.org/wiki/Wealth_in_the_United-ed_States. Since I used individuals and not families in the example with the bus, I took 50 percent of that figure. That's not a correct figure, since individuals who live by themselves also constitute a household in the technical sense. But the exact number doesn't matter for the example.

MOTIVATION CROWDING

Bruno S. Frey, "Die Grenzen ökonomischer Anreize," *Neue Zürcher Zeitung*, May 18, 2001. (Translation: "The Limits of Economic Incentives." Bruno Frey makes the case to scientifically study intrinsic motivation instead of [mostly] monetary incentives. There is no English translation of this article.)

This paper provides a good overview: Bruno S. Frey and Reto Jegen, "Motivation Crowding Theory: A Survey of Empirical Evidence," *Journal of Economic Surveys* 15, no. 5 (2001): 589–611.

The story with the day care center: Steven D. Levitt and Stephen J. Dubner, *Freakonomics: A Rogue Economist Explores the Hidden Side of Everything* (New York: William Morrow, 2005), 19.

Ori Brafman and Rom Brafman, *Sway: The Irresistible Pull of Irrational Behavior* (New York: Doubleday, 2008), 131–35.

It's not all black and white. In certain settings, pay for performance can also have a positive effect on self-determination and task enjoyment. Robert Eisenberger, Linda Rhoades, and Judy Cameron, "Does Pay for Performance Increase or Decrease Perceived Self-Determination and Intrinsic

Motivation?," *Journal of Personality and Social Psychology* 77, no. 5 (1999): 1026–40.

There are so many examples of *motivation crowding*, and the scientific literature is ample. Here is an example: "Every year, on a predetermined day, students go from house to house collecting monetary donations that households make to societies for cancer research, help for disabled children, and the like. Students performing these activities typically receive much social approval from parents, teachers, and other people. This is the very reason why they perform these activities voluntarily. When students were each offered one percent of the money they collected, the amount collected decreased by 36 percent." Ernst Fehr and Armin Falk, "Psychological Foundations of Incentives," *European Economic Review* 46 (May 2002): 687–724.

TWADDLE TENDENCY

An example of smoke screen writing: Jürgen Habermas, *Between Facts and Norms: Contributions to a Discourse Theory of Law and Democracy* (Cambridge, MA: MIT Press, 1998), 490.

WILL ROGERS PHENOMENON

Stage migration when diagnosing tumors goes even further than described in the chapter. Because stage 1 now contains so many cases, doctors adjust the boundaries between stages. The worst stage 1 patients are categorized as stage 2, the worst stage 2 patients as stage 3, and the worst stage 3 patients as stage 4. Each of these new additions raises the average life expectancy of the group. The result: Not a single patient lives longer. It appears that the therapy has helped patients, but merely the diagnosis has improved. A. R. Feinstein, D. M. Sosin, and C. K. Wells, "The Will Rogers Phenomenon—Stage Migration and New Diagnostic Techniques as a Source of Misleading Statistics for Survival in Cancer," *New England Journal of Medicine* 312, no. 25 (June 1985): 1604–8.

Further examples can be found in the excellent book: Hans-Hermann Dubben and Hans-Peter Beck-Bornholdt, *Der Hund, der Eier legt: Erkennen von Fehlinformation durch Querdenken* (Reinbek, Germany: Rororo Publisher, 2006), 34–235. There is no English translation of this book.

INFORMATION BIAS

"To bankrupt a fool, give him information." Nassim Nicholas Taleb, *The Bed of Procrustes: Philosophical and Practical Aphorisms* (New York: Random House, 2010), 4.

The example with the three diseases: Jonathan Baron, Jane Beattie, and John C. Hershey, "Heuristics and Biases in Diagnostic Reasoning: II. Congruence, Information, and Certainty," *Organizational Behavior and Human Decision Processes* 42 (1988): 88–110.

EFFORT JUSTIFICATION

For Aronson and Mills the effort justification is nothing but the reduction of cognitive dissonance. Elliot Aronson and Judson Mills, "The Effect of Severity of Initiation on Liking for a Group," *Journal of Abnormal and Social Psychology* 59 (1959): 177–81.

Michael I. Norton: Michael I. Norton, Daniel Mochon, and Dan Ariely, "The IKEA Effect: When Labor Leads to Love," *Journal of Consumer Psychology* 22, no. 3 (July 2012): 453–60.

THE LAW OF SMALL NUMBERS

Daniel Kahneman uses a good example in his book *Thinking, Fast and Slow* (New York: Farrar, Straus and Giroux, 2011), 109–113. My story with the shoplifting rates borrows heavily from this.

EXPECTATIONS

In the main text, we did not cover asymmetry. Shares that exceed expectations rise, on average, by 1 percent. Shares that fall below expectations drop, on average, by 3.4 percent. See: Jason Zweig, *Your Money and Your Brain* (New York: Simon & Schuster, 2007), 181.

Rosenthal effect: Robert Rosenthal and Leonore Jacobson, *Pygmalion in the Classroom*, expanded ed. (New York: Irvington, 1968).

Robert S. Feldman and Thomas Prohaska, "The Student as Pygmalion: Effect of Student Expectation on the Teacher," *Journal of Educational Psychology* 71, no. 4 (1979): 485–93.

SIMPLE LOGIC

The original paper on the CRT: Shane Frederick, "Cognitive Reflection and Decision Making," *Journal of Economic Perspectives* 19, no. 4 (Fall 2005): 25–42.

Amitai Shenhav, David G. Rand, and Joshua D. Greene, "Divine Intuition: Cognitive Style Influences Belief in God," *Journal of Experimental Psychology* 141, no. 3 (August 2012): 423–28.

FORER EFFECT

Bertram R. Forer, "The Fallacy of Personal Validation: A Classroom Demonstration of Gullibility," *Journal of Abnormal and Social Psychology* 44, no. 1 (1949): 118–23.

This is also called the "Barnum effect." Ringmaster Phineas T. Barnum designed his show around the motto: "a little something for everybody."

Joel T. Johnson, Lorraine M. Cain, Toni L. Falke, Jon Hayman, and Edward Perillo, "The 'Barnum Effect' Revisited: Cognitive and Motivational Factors in the Acceptance of Personality Descriptions," *Journal of Personality and Social Psychology* 49, no. 5 (November 1985): 1378–91.

D. H. Dickson and I. W. Kelly, "The 'Barnum Effect' in Personality Assessment: A Review of the Literature," *Psychological Reports* 57 (1985): 367–82.

The Skeptic's Dictionary has a good entry on the Forer Effect: http://www.skepdic.com/forer.html.

VOLUNTEER'S FOLLY

No topic has drawn more feedback than this (previously these chapters were newspaper columns). One reader commented that it would be even better to have the birdhouses manufactured in China than to get a local carpenter to make them. The reader is right, of course, providing you subtract the environmental damage caused by the shipping. The point is that *volunteer's folly* is nothing more than David Ricardo's law of comparative advantage.

Trevor M. Knox, "The Volunteer's Folly and Socio-Economic Man: Some Thoughts on Altruism, Rationality, and Community," *Journal of Socio-Economics* 28, no. 4 (1999): 475–92.

AFFECT HEURISTIC

Daniel Kahneman, *Thinking, Fast and Slow* (New York: Farrar, Straus and Giroux, 2011), 139–42.

Priming the affect through smilies or frownies before judging Chinese icons: Sheila T. Murphy, Jennifer L. Monahan, and R. B. Zajonc, "Additivity of Nonconscious Affect: Combined Effects of Priming and Exposure," *Journal of Personality and Social Psychology* 69, no. 4 (October 1995): 589–602.

See also: Piotr Winkielman, Robert B. Zajonc, and Norbert Schwarz, "Subliminal Affective Priming Attributional Interventions," *Cognition and Emotion* 11, no. 4 (1997): 433–65.

How morning sun affects the stock market: David Hirshleifer and Tyler Shumway, "Good Day Sunshine: Stock Returns and the Weather," *Journal of Finance* 58, no. 3 (2003): 1009–32.

INTROSPECTION ILLUSION

Kathryn Schulz, *Being Wrong: Adventures in the Margin of Error* (New York: Ecco, 2010), 104–10. I've adapted Schulz's green teas story and made it into a story of a vitamin pill producer.

Much of the introspection illusion comes down to "shallow thinking": Thomas Gilovich, Nicholas Epley, and Karlene Hanko, "Shallow Thoughts about the Self: The Automatic Components of Self-Assessment," in Mark D. Alicke, David A. Dunning, and Joachim I. Krueger, *The Self in Social Judgment: Studies in Self and Identity* (New York: Psychology Press, 2005), 67–81.

Richard E. Nisbett and Timothy D. Wilson, "Telling More Than We Can Know: Verbal Reports on Mental Processes," *Psychological Review* 84 (1977): 231–59.

INABILITY TO CLOSE DOORS

Dan Ariely, *Predictably Irrational: The Hidden Forces That Shape Our Decisions*, rev. and expanded ed. (New York: HarperCollins, 2008), chapter 9, "Keeping Doors Open," 183–98.

Mark Edmundson describing today's generation of students: "They want to study, travel, make friends, make more friends, read everything (superfast), take in all the movies, listen to every hot band, keep up with everyone they've ever known. And there's something else, too, that distinguishes them: They live to multiply possibilities. They're enemies of closure. For as much as they want to do and actually manage to do, they always strive to keep their options open, never to shut possibilities down before they have to." Mark Edmundson, "Dwelling in Possibilities," *Chronicle of Higher Education*, March 14, 2008.

NEOMANIA

Nassim Nicholas Taleb, *Antifragile: Things That Gain from Disorder* (New York: Random House, 2012), 322–28.

SLEEPER EFFECT

Carl Hovland carried out his tests using the propaganda movie *Why We Fight*. The movie is available on YouTube.

See also: Gareth Cook, "TV's Sleeper Effect: Misinformation on Television Gains Power over Time," *Boston Globe*, October 30, 2011.

Beliefs acquired by reading fictional narratives are integrated into real-world knowledge. In: Markus Appel and Tobias Richter, "Persuasive Effects of Fictional Narratives Increase over Time," *Media Psychology* 10 (2007): 113–34.

Tarcan G. Kumkale and Dolore Albarracín, "The Sleeper Effect in Persuasion: A Meta-Analytic Review," *Psychological Bulletin* 130, no. 1 (January 2004): 143–72.

David Mazursky and Yaacov Schul, "The Effects of Advertisement Encoding on the Failure to Discount Information: Implications for the Sleeper Effect," *Journal of Consumer Research* 15, no. 1 (1988): 24–36.

Ruth Ann Weaver Lariscy and Spencer F. Tinkham, "The Sleeper Effect and Negative Political Advertising," *Journal of Advertising* 28, no. 4 (Winter 1999): 13–30.

SOCIAL COMPARISON BIAS

Stephen M. Garcia, Hyunjin Song, and Abraham Tesser, "Tainted Recommendations: The Social Comparison Bias," *Organizational Behavior and Human Decision Processes* 113, no. 2 (2010): 97–101.

B-players hire C-players, and so on. Watch this excellent video on YouTube: Guy Kawasaki, *The Art of the Start*.

By the way: Some authors succeed at mutually flattering each another, such as Niall Ferguson and Ian Morris. They continually bestow the title of "best historian" upon each other. Clever. It's rare, a perfected art.

PRIMACY AND RECENCY EFFECTS

Primacy effect: Psychologist Solomon Asch scientifically investigated this in the 1940s. The example using Alan und Ben comes from him. Solomon E. Asch, "Forming Impressions of Personality," *Journal of Abnormal and Social Psychology* 41, no. 3 (July 1946): 258–90.

The example from Alan and Ben cited in: Daniel Kahneman, *Thinking, Fast and Slow* (New York: Farrar, Straus and Giroux, 2011), 82–83.

The final ad before a film starts is the most expensive for another reason: It will reach the most people since everyone will have taken their seats by then.

There is a myriad of research on the *primacy and recency effects*. Here are two papers: Arthur M. Glenberg et al., "A Two-Process Account of Long-Term Serial Position Effects," *Journal of Experimental Psychology: Human*

Learning and Memory 6, no. 4 (July 1980): 355–69. And: M. W. Howard and M. Kahana, "Contextual Variability and Serial Position Effects in Free Recall," *Journal of Experimental Psychology: Learning, Memory and Cognition* 25, no. 4 (July 1999): 923–41.

NOT-INVENTED-HERE SYNDROME

Ralph Katz and Thomas J. Allen, "Investigating the Not Invented Here (NIH) Syndrome: A Look at the Performance, Tenure and Communication Patterns of 50 R&D Project Groups," *R&D Management* 12, no. 1 (1982): 7–19.

Joel Spolsky wrote an interesting blog entry contesting NIH syndrome. It's available online under the name: *In Defense of Not-Invented-Here Syndrome* (in http://www.joelonsoftware.com, October 14, 2001). His theory: World-class teams should not be dependent on the developments of other teams or other companies. When developing any in-house product, you should design the central part yourself from top to bottom. This reduces dependencies and guarantees the highest quality.

THE BLACK SWAN

Nassim Nicholas Taleb, *The Black Swan: The Impact of the Highly Improbable* (New York: Random House, 2007).

DOMAIN DEPENDENCE

"Upon arriving at the hotel in Dubai, the businessman had a porter carry his luggage; I later saw him lifting free weights in the gym." Nassim Nicholas Taleb, *The Bed of Procrustes: Philosophical and Practical Aphorisms* (New York: Random House, 2010), 75.

Another brilliant aphorism by Taleb on the subject: "My best example of domain dependence of our minds, from my recent visit to Paris: at lunch in a French restaurant, my friends ate the salmon and threw away the skin; at dinner, at the sushi bar, the very same friends ate the skin and threw away the salmon." Ibid., 76.

Domestic violence is two to four times more common in police families than in the general population. Read: Peter H. Neidig, Harold E. Russell, and Albert F. Seng, "Interspousal Aggression in Law Enforcement Families: A Preliminary Investigation," *Police Studies* 15, no. 1 (1992): 30–38.

L. D. Lott, "Deadly Secrets: Violence in the Police Family," *FBI Law Enforcement Bulletin* 64 (November 1995): 12–16.

The Markowitz example: "I should have computed the historical covariance of the asset classes and drawn an efficient frontier. Instead I visualized my grief if the stock market went way up and I wasn't in it—or if it went way down and I was completely in it. My intention was to minimize my future regret, so I split my [pension scheme] contributions 50/50 between bonds and equities." Harry Markowitz, quoted in Jason Zweig, "How the Big Brains Invest at TIAA-CREF," *Money* 27, no. 1 (January 1998): 114. See also: Jason Zweig, *Your Money and Your Brain* (New York: Simon & Schuster, 2007), 4.

The Bobbi Bensman example: Zweig, *Your Money and Your Brain*, 127.

Domain specificity is connected to the modular structure of the brain. If you are skilled with your hands (like pianists), it does not mean that you will have equally reactive legs (like footballers). Though both brain regions are in the "motor cortex," they are not in the same place—they are not even next to each other.

The quote from Barry Mazur see: Barry C. Mazur, presentation given at 1865th Stated Meeting titled *The Problem of Thinking Too Much*, December 11, 2002, http://www.amacad.org/publications/bulletin/spring2003/diaconis.pdf.

FALSE-CONSENSUS EFFECT

Thomas Gilovich, Dale Griffin, and Daniel Kahneman (eds.), *Heuristics and Biases: The Psychology of Intuitive Judgment* (Cambridge, UK: Cambridge University Press, 2002), 642.

The sandwich board "Eat at Joe's" example: Lee Ross, David Greene, and Pamela House, "The 'False Consensus Effect': An Egocentric Bias in Social Perception and Attribution Processes," *Journal of Personality and Social Psychology* 13, no. 3 (May 1977): 279–301.

This effect overlaps with other mental errors. For example, the *availability bias* can lead into the *false consensus effect*. Whoever deliberates on a question can easily recall their conclusions (they are available). The person wrongly assumes that these findings will be as readily available to someone else. The *self-serving bias* also influences the *false-consensus effect*. Whoever wants to present something in a convincing manner does well to tell themselves that many (maybe even the majority) share their view and that their ideas will not fall on deaf ears. Philosophy deems the *false-consensus effect* "naive realism": People are convinced that their positions are well thought out. Whoever fails to share their views will see the light if they reflect and open their minds sufficiently.

The *false-consensus effect* can be reduced by explaining or showing subjects both sides of the story. Kathleen P. Bauman and Glenn Geher, "We Think

You Agree: The Detrimental Impact of the False Consensus Effect on
Behavior," *Current Psychology* 21, no. 4 (2002): 293–318.

FALSIFICATION OF HISTORY
More information on Gregory Markus: See: Kathryn Schulz, *Being Wrong:
Adventures in the Margin of Error* (New York: Ecco, 2010), 185.
Gregory Markus, "Stability and Change in Political Attitudes: Observe,
Recall and Explain," *Political Behavior* 8 (1986): 21–44.
Flashbulb memory: Ibid., 17–73.
In 1902, University of Berlin criminology professor Franz von Liszt (nothing
to do with the composer Franz Liszt) showed that the best witnesses in
court recall at least a fourth of the facts incorrectly. Ibid., 223.

IN-GROUP OUT-GROUP BIAS
"Life in nature involves competition, and groups can certainly compete
better than individuals. The hidden dimension is that individuals cannot
usually compete against groups. Therefore, once groups exist anywhere,
everyone else has to join a group, if only for self-protection." Roy F.
Baumeister, *The Cultural Animal: Human Nature, Meaning, and Social Life*
(Oxford, UK: Oxford University Press, 2005), 377–79.
The classic paper: Henri Tajfel, "Experiments in Intergroup Discrimination,"
Scientific American 223 (1970): 96–102.
For agreement surplus in groups, see: Kathryn Schulz, *Being Wrong: Adventures in the Margin of Error* (New York: Ecco, 2010), 149.
More about "pseudokinship," see Robert Sapolsky, "Anthropology/Humans
Can't Smell Trouble/"Pseudokinship" and Real War," *SF Gate*, March 2,
2003, http://www.sfgate.com/opinion/article/ANTHROPOLOGY-
Humans-Can-t-Smell-Trouble–2666430.php.

AMBIGUITY AVERSION
Knightian uncertainty is named after University of Chicago economist Frank
Knight (1885–1972), who distinguished risk and uncertainty in his work:
Frank H. Knight, *Risk, Uncertainty, and Profit* (Boston: Houghton Mifflin
Company, 1921).
The Ellsberg paradox is actually a little more complicated. A detailed expla-
nation is available on Wikipedia (http://en.wikipedia.org/wiki/
Ellsberg_paradox).
Yes, we curse uncertainty. But it has its positive sides. Suppose you live in a
dictatorship and want to get past the censors. You can resort to ambiguity.

DEFAULT EFFECT

The car insurance policies: Jonathan Baron, *Thinking and Deciding* (Cambridge, UK: Cambridge University Press, 2000), 299.

Eric J. Johnson and Daniel Goldstein, "Do Defaults Save Lives?," *Science* 302, no. 5649 (November 2003): 1338–39.

Cass Sunstein and Richard Thaler, *Nudge: Improving Decisions about Health, Wealth, and Happiness* (New Haven, CT: Yale University Press, 2008).

The difficulties of renegotiating contracts: Daniel Kahneman, *Thinking, Fast and Slow* (New York: Farrar, Straus and Giroux, 2011), 304–5.

FEAR OF REGRET

The story with Paul and George: Daniel Kahneman and Amos Tversky, "Intuitive Prediction: Biases and Corrective Procedures," in Daniel Kahneman, Paul Slovic, and Amos Tversky, *Judgment under Uncertainty: Heuristics and Biases* (New York: Cambridge University Press, 1982), 414–21.

The passenger who should not have been on the plane that crashed: Daniel Kahneman, *Thinking, Fast and Slow* (New York: Farrar, Straus and Giroux, 2011), 346–48.

For traders' off-loading, see: Meir Statman and Kenneth L. Fisher, "Hedging Currencies with Hindsight and Regret," *Journal of Investing* 14 (2005): 15–19.

Ilana Ritov and Jonathan Baron, "Outcome Knowledge, Regret, and Omission Bias," *Organizational Behavior and Human Decision Processes* 64 (1995): 119–27.

Another regret question is the following: On your way to the airport you are caught in a traffic jam. You arrive at the airport thirty minutes after scheduled departure time. What makes you more upset (more regret): (a) your flight left on time, (b) your flight was delayed and it left only five minutes ago. Most people answer with (b). The example is again from Kahneman and Tversky. I shortened it a bit. The original wording in: Daniel Kahneman and Amos Tversky, "The Psychology of Preferences," *Scientific American* 246 (1982): 160–73.

An example of fear of regret. " 'A Fear of Regret Has Always Been My Inspiration': Maurizio Cattelan on His Guggenheim Survey," *Blouin ArtInfo*, November 2, 2011.

We empathize more with Anne Frank than with a similar girl who was immediately arrested and sent to Auschwitz. Compared to other detentions, Anne Frank's is an exception. Of course, the availability bias also plays a role. Anne Frank's story is known worldwide through her diary. Most other detentions are forgotten and therefore not available to us.

A Note on Sources

SALIENCE EFFECT

Roy F. Baumeister, *The Cultural Animal: Human Nature, Meaning, and Social Life* (Oxford, UK: Oxford University Press, 2005), 211.

Werner F. M. De Bondt and Richard H. Thaler, "Do Analysts Overreact?," in Thomas Gilovich, Dale Griffin, and Daniel Kahneman (eds.), *Heuristics and Biases: The Psychology of Intuitive Judgment* (Cambridge, UK: Cambridge University Press, 2002), 678–85.

Scott Plous, *The Psychology of Judgment and Decision Making* (New York: McGraw-Hill, 1993), 125–27. Plous substitutes "salience" with "vividness." The two are similar.

The *salience effect* is related to the *availability bias*. With both effects, information that is more easily accessible enjoys undue explanatory power or leads to above-average motivation.

HOUSE-MONEY EFFECT

Cass Sunstein and Richard Thaler, *Nudge: Improving Decisions about Health, Wealth, and Happiness* (New Haven, CT: Yale University Press, 2008), 54–55.

Peter L. Bernstein, *Against the Gods: The Remarkable Story of Risk* (New York: Wiley, 1996), 274–75.

You've just received: Carrie M. Heilman, Kent Nakamoto, and Ambar G. Rao, "Pleasant Surprises: Consumer Response to Unexpected In-Store Coupons," *Journal of Marketing Research* 39, no. 2 (May 2002): 242–52.

Pamela W. Henderson and Robert A. Peterson, "Mental Accounting and Categorization," *Organizational Behavior and Human Decision Processes* 51, no. 1 (February 1992): 92–117.

The government can utilize the *house-money effect*. As part of President Bush's 2001 tax reform, each American taxpayer received a credit of $600. People who viewed this as a gift from the government spent more than three times as much as those who saw it as their own money. In this way, tax credits can be used to stimulate the economy.

PROCRASTINATION

Jason Zweig, *Your Money and Your Brain* (New York: Simon & Schuster, 2007), 253–54.

On the effectiveness of self-imposed deadlines: Dan Ariely and Klaus Wertenbroch, "Procrastination, Deadlines, and Performance: Self-Control by Precommitment," *Psychological Science* 13, no. 3 (May 1, 2002): 219–24.

ENVY

Envy is one of the Catholic Church's seven deadly sins. In the book of
Genesis, Cain kills his brother Abel out of envy because God prefers his
sacrifice. This is the first murder in the Bible.

One of the floweriest accounts of envy is the fairy tale "Snow White and
the Seven Dwarves." In the story, Snow White's stepmother envies her
beauty. First, she hires an assassin to kill her, but he does not go through
with it. Snow White flees into the forest to the seven dwarfs. Outsourcing
didn't work so well, so now the stepmother has to take matters into her
own hands. She poisons the beautiful Snow White.

Munger: "The idea of caring that someone is making money faster than you
are is one of the deadly sins. Envy is a really stupid sin because it's the
only one you could never possibly have any fun at. There's a lot of pain
and no fun. Why would you want to get on that trolley?" in Charles T.
Munger, *Poor Charlie's Almanack*, expanded 3rd ed. (Virginia Beach, VA:
The Donning Company Publishers, 2006), 138.

Of course, not all envy is spiteful—there are also innocent episodes, such as
a grandfather envying his grandchildren's youth. This is not resentment;
the older man would simply like to be young and carefree again.

PERSONIFICATION

Deborah A. Small, George Loewenstein, and Paul Slovic, "Sympathy and
Callousness: The Impact of Deliberative Thought on Donations to
Identifiable and Statistical Victims," *Organizational Behavior and Human
Decision Processes* 102, no. 2 (2007): 143–53.

"If I look at the mass, I will never act. If I look at the one, I will." Mother
Teresa in ibid.

ILLUSION OF ATTENTION

Christopher Chabris and Daniel Simons, *The Invisible Gorilla: And Other
Ways Our Intuitions Deceive Us* (New York: Crown, 2010), 1–42.

For using your cell phone while driving, see: Donald D. Redelmeier and
Robert J. Tibishirani, "Association between Cellular-Telephone Calls and
Motor Vehicle Collisions," *New England Journal of Medicine* 336 (1997):
453–58.

See also: David L. Strayer, Frank A. Drews, and Dennis J. Crouch, "Comparing the Cell-Phone Driver and the Drunk Driver," *Human Factors* 48
(2006): 381–91.

And, if instead of phoning someone, you chat with whomever is in the

passenger seat? Research found no negative effects. First, face-to-face
conversations are much clearer than phone conversations, that is, your
brain must not work so hard to decipher the messages. Second, your
passenger understands that if the situation gets dangerous, the chatting
will be interrupted. That means you do not feel compelled to continue the
conversation. Third, your passenger has an additional pair of eyes and can
point out dangers.

STRATEGIC MISREPRESENTATION

Flyvbjerg defines *strategic misrepresentation* as "lying, with a view to getting
projects started." Bent Flyvbjerg, *Megaprojects and Risk: An Anatomy of
Ambition* (Cambridge, UK: Cambridge University Press, 2003), 16.

L. R. Jones and K. J. Euske, "Strategic Misrepresentation in Budgeting,"
Journal of Public Administration Research and Theory 1, no. 4 (October
1991): 437–60.

In online dating, men are more likely to misrepresent personal assets, relationship goals, personal interests, and personal attributes, whereas women
are more likely to misrepresent weight: Jeffrey A. Hall et al., "Strategic
Misrepresentation in Online Dating," *Journal of Social and Personal Rela-
tionships* 27, no. 1 (2010): 117–35.

OVERTHINKING

Timothy D. Wilson and Jonathan W. Schooler, "Thinking Too Much: Introspection Can Reduce the Quality of Preferences and Decisions," *Journal
of Personality and Social Psychology* 60, no. 2 (February 1991): 181–92.

Known to chess players as the Kotov syndrome: A player contemplates too
many moves, fails to come to a decision, and, under time pressure, makes
a rookie mistake.

PLANNING FALLACY

Roger Buehler, Dale Griffin, and Michael Ross, "Inside the Planning Fallacy: The Causes and Consequences of Optimistic Time Predictions," in
Thomas Gilovich, Dale Griffin, and Daniel Kahneman (eds.), *Heuristics
and Biases: The Psychology of Intuitive Judgment* (Cambridge, UK: Cambridge University Press, 2002), 250–70.

Gary Klein doesn't spell out the exact speech as mentioned in this chapter.
This is how he prescribes it: "A typical premortem begins after the team
has been briefed on the plan. The leader starts the exercise by inform-

ing everyone that the project has failed spectacularly. Over the next few minutes those in the room independently write down every reason they can think of for the failure—especially the kinds of things they ordinarily wouldn't mention as potential problems, for fear of being impolitic." See: Gary Klein, "Performing a Project Premortem," *Harvard Business Review*, *http://hbr.org/2007/09/performing-a-project-premortem/ar/1*. Accessed December 17, 2012.

Samuel Johnson wrote: People who remarry represent "the triumph of hope over experience"—in James Boswell's *Life of Samuel Johnson* (London: Printed by Henry Baldwin for Charles Dilly, in the Poultry, 1791). In making plans, we are all serial brides and grooms.

Hofstadter's Law: "It always takes longer than you expect, even when you take into account Hofstadter's Law." Douglas Hofstadter, *Gödel, Escher, Bach: An Eternal Golden Braid*, 20th anniversary ed. (New York: Basic Books, 1999), 152.

The *planning fallacy* is related to the *overconfidence effect*. With the *overconfidence effect*, we believe our capabilities are greater than they are, whereas the *planning fallacy* leads us to overestimate our abilities, turnaround times, and budgets. In both cases, we are convinced that the error rate of our predictions (whether in terms of achieving goals or forecasting timelines) is smaller than it actually is. In other words, we know we make mistakes when estimating durations. But we are confident that they will happen only rarely or not at all.

A great example of a premortem is described in: Daniel Kahneman, *Thinking, Fast and Slow* (New York: Farrar, Straus and Giroux, 2011), 264.

The Danish planning expert Bent Flyvbjerg has researched mega-projects more than anyone else. His conclusion: "The prevalent tendency to underweight distributional information is perhaps the major source of error in forecasting." Quoted in ibid., 251.

The *planning fallacy* in the military: "No battle plan survives contact with the enemy." The saying is attributed to German military strategist Helmuth von Moltke.

See also: Roy F. Baumeister, *The Cultural Animal: Human Nature, Meaning, and Social Life* (Oxford, UK: Oxford University Press, 2005), 241–44.

Here's a great way to avoid the *planning fallacy* even if you don't have access to a database of similar projects: "You can ask other people to take a fresh look at your ideas and make their own forecast for the project. Not a forecast of how long it would take *them* to execute the ideas (since they too will likely underestimate their own time and costs), but of how long it will take *you* (or your contractors, employees, etc.) to do so." Quoted from Christopher Chabris and Daniel Simons, *The Invisible Gorilla: And Other Ways Our Intuitions Deceive Us* (New York: Crown, 2010), 127.

DÉFORMATION PROFESSIONNELLE

"You've got to have models across a wide array of disciplines." Charles T. Munger, *Poor Charlie's Almanack*, expanded 3rd ed. (Virginia Beach, VA: The Donning Company Publishers, 2006), 167.

ZEIGARNIK EFFECT

Roy Baumeister and John Tierney, *Willpower: Rediscovering the Greatest Human Strength* (New York: Penguin Press, 2011), 80–82.

Whether it was a scarf or something else that was left in the restaurant we do not know. We also do not know if it was Bluma Zeigarnik who went back to the restaurant. To make the chapter more fluid, I assumed these were the case.

ILLUSION OF SKILL

Daniel Kahneman, *Thinking, Fast and Slow* (New York: Farrar, Straus and Giroux, 2011), 204–21.

Warren Buffett: "My conclusion from my own experiences and from much observation of other businesses is that a good managerial record (measured by economic returns) is far more a function of what business boat you get into than it is of how effectively you row (though intelligence and effort help considerably, of course, in any business, good or bad). Some years ago I wrote: 'When a management with a reputation for brilliance tackles a business with a reputation for poor fundamental economics, it is the reputation of the business that remains intact.' Nothing has since changed my point of view on that matter." Warren Buffett, letter to shareholders of Berkshire Hathaway, 1985.

FEATURE-POSITIVE EFFECT

The antismoking campaign: Guangzhi Zhao and Cornelia Pechmann, "Regulatory Focus, Feature Positive Effect, and Message Framing," *Advances in Consumer Research* 33, no. 1 (2006): 100.

An overview of the research on the *feature-positive effect*: Frank R. Kardes, David M. Sanbonmatsu, and Paul M. Herr, "Consumer Expertise and the Feature-Positive Effect: Implications for Judgment and Inference," *Advances in Consumer Research* 17 (1990): 351–54.

CHERRY PICKING

"The harmful effects of smoking are roughly equivalent to the combined good ones of every medical intervention developed since the war. Those who smoke, in other words, now have the same life expectancy as if they were non-smokers without access to any health care developed in the last half-century. Getting rid of smoking provides more benefit than being able to cure people of every possible type of cancer." Druin Burch, *Taking the Medicine: A Short History of Medicine's Beautiful Idea and Our Difficulty Swallowing It* (London: Chatto & Windus, 2009), 238.

Cherry picking in religion: People take what suits them from the Bible and ignore the other teachings. If we wanted to follow the Bible literally, we would have to stone disobedient sons and unfaithful wives (Deuteronomy 21 and 22) and kill all homosexuals (Leviticus 20:13).

Cherry picking in forecasting: Forecasts that turn out to be correct are announced triumphantly. Wrong prognoses remain "unpicked." See the chapter on the *forecast illusion*.

FALLACY OF THE SINGLE CAUSE

Chris Matthews cited in: Christopher Chabris and Daniel Simons, *The Invisible Gorilla: And Other Ways Our Intuitions Deceive Us* (New York: Crown, 2010), 172. The authors highlighted the quotes.

Leo Tolstoy, *War and Peace* (New York: Vintage Classics, 2008), 606.

A great essay on the *fallacy of the single cause*: John Tooby, "Nexus Causality, Moral Warfare, and Misattribution Arbitrage," in John Brockman, *This Will Make You Smarter* (New York: Harper, 2012), 34–35.

INTENTION-TO-TREAT ERROR

Hans-Hermann Dubben and Hans-Peter Beck-Bornholdt, *Der Hund, der Eier legt: Erkennen von Fehlinformation durch Querdenken* (Reinbek, Germany: Rororo Publishers, 2006), 238–39. Unfortunately, no English translation of this excellent book exists.

For a full description of the *intention-to-treat error*, sometimes also referred to as "intent-to-treat," read: John M. Lachin, "Statistical Considerations in the Intent-to-Treat Principle," *Controlled Clinical Trials* 21, no. 5 (October 2000): 526.

EPILOGUE

Via Negativa: "Charlie generally focuses first on what to avoid—that is, on what NOT to do—before he considers the affirmative steps he will take in a given situation. 'All I want to know is where I'm going to die, so I'll never go there' is one of his favorite quips." In: Charles T. Munger, *Poor Charlie's Almanack*, expanded 3rd ed. (Virginia Beach, VA: The Donning Company Publishers, 2006), 63.

Via Negativa: "Part of (having uncommon sense) is being able to tune out folly, as opposed to recognizing wisdom." Ibid., 134.

ABOUT THE AUTHOR

ROLF DOBELLI, born in 1966, is a Swiss novelist. This is his first work of nonfiction. He earned his MBA from the University of St. Gallen, Switzerland, and received his PhD for a dissertation in philosophy. He is the founder or co-founder of several companies and communities, including: ZURICH .MINDS, a community of the leading personalities from science, culture, and business; and getAbstract, the world's largest resource of compressed business literature. Rolf Dobelli lives in Lucerne, Switzerland.

Visit the author's website: www.rolfdobelli.com.